192
Advances in Polymer Science

Editorial Board:
A. Abe · A.-C. Albertsson · R. Duncan · K. Dušek · W. H. de Jeu
J.-F. Joanny · H.-H. Kausch · S. Kobayashi · K.-S. Lee · L. Leibler
T. E. Long · I. Manners · M. Möller · O. Nuyken · E. M. Terentjev
B. Voit · G. Wegner · U. Wiesner

Advances in Polymer Science

Recently Published and Forthcoming Volumes

Surface-Initiated Polymerization II
Volume Editor: Jordan, R.
Vol. 198, 2006

Surface-Initiated Polymerization I
Volume Editor: Jordan, R.
Vol. 197, 2006

Conformation-Dependent Design of Sequences in Copolymers II
Volume Editor: Khokhlov, A. R.
Vol. 196, 2006

Conformation-Dependent Design of Sequences in Copolymers I
Volume Editor: Khokhlov, A. R.
Vol. 195, 2006

Enzyme-Catalyzed Synthesis of Polymers
Volume Editors: Kobayashi, S., Ritter, H., Kaplan, D.
Vol. 194, 2006

Polymer Therapeutics II
Polymers as Drugs, Conjugates and Gene Delivery Systems
Volume Editors: Satchi-Fainaro, R., Duncan, R.
Vol. 193, 2006

Polymer Therapeutics I
Polymers as Drugs, Conjugates and Gene Delivery Systems
Volume Editors: Satchi-Fainaro, R., Duncan, R.
Vol. 192, 2006

Interphases and Mesophases in Polymer Crystallization III
Volume Editor: Allegra, G.
Vol. 191, 2005

Block Copolymers II
Volume Editor: Abetz, V.
Vol. 190, 2005

Block Copolymers I
Volume Editor: Abetz, V.
Vol. 189, 2005

Intrinsic Molecular Mobility and Toughness of Polymers II
Volume Editor: Kausch, H.-H.
Vol. 188, 2005

Intrinsic Molecular Mobility and Toughness of Polymers I
Volume Editor: Kausch, H.-H.
Vol. 187, 2005

Polysaccharides I
Structure, Characterization and Use
Volume Editor: Heinze, T.
Vol. 186, 2005

Advanced Computer Simulation Approaches for Soft Matter Sciences II
Volume Editors: Holm, C., Kremer, K.
Vol. 185, 2005

Crosslinking in Materials Science
Vol. 184, 2005

Phase Behavior of Polymer Blends
Volume Editor: Freed, K.
Vol. 183, 2005

Polymer Analysis/Polymer Theory
Vol. 182, 2005

Interphases and Mesophases in Polymer Crystallization II
Volume Editor: Allegra, G.
Vol. 181, 2005

Interphases and Mesophases in Polymer Crystallization I
Volume Editor: Allegra, G.
Vol. 180, 2005

Polymer Therapeutics I

Polymers as Drugs, Conjugates and Gene Delivery Systems

Volume Editors: Ronit Satchi-Fainaro · Ruth Duncan

With contributions by

R. J. Amir · P. K. Dhal · R. Duncan · S. R. Holmes-Farley
C. C. Huval · T. H. Jozefiak · J. Kloeckner · G. Pasut
H. Ringsdorf · R. Satchi-Fainaro · D. Shabat
F. M. Veronese · E. Wagner

The series *Advances in Polymer Science* presents critical reviews of the present and future trends in polymer and biopolymer science including chemistry, physical chemistry, physics and material science. It is adressed to all scientists at universities and in industry who wish to keep abreast of advances in the topics covered.

As a rule, contributions are specially commissioned. The editors and publishers will, however, always be pleased to receive suggestions and supplementary information. Papers are accepted for *Advances in Polymer Science* in English.

In references *Advances in Polymer Science* is abbreviated *Adv Polym Sci* and is cited as a journal.

Springer WWW home page: http://www.springer.com
Visit the APS content at http://www.springerlink.com/

Library of Congress Control Number: 2005933608

ISSN 0065-3195
ISBN-10 3-540-29210-1 Springer Berlin Heidelberg New York
ISBN-13 978-3-540-29210-4 Springer Berlin Heidelberg New York
DOI 10.1007/11547761

This work is subject to copyright. All rights are reserved, whether the whole or part of the material is concerned, specifically the rights of translation, reprinting, reuse of illustrations, recitation, broadcasting, reproduction on microfilm or in any other way, and storage in data banks. Duplication of this publication or parts thereof is permitted only under the provisions of the German Copyright Law of September 9, 1965, in its current version, and permission for use must always be obtained from Springer. Violations are liable for prosecution under the German Copyright Law.

Springer is a part of Springer Science+Business Media

springer.com

© Springer-Verlag Berlin Heidelberg 2006
Printed in Germany

The use of registered names, trademarks, etc. in this publication does not imply, even in the absence of a specific statement, that such names are exempt from the relevant protective laws and regulations and therefore free for general use.

Cover design: *Design & Production* GmbH, Heidelberg
Typesetting and Production: LE-TEX Jelonek, Schmidt & Vöckler GbR, Leipzig

Printed on acid-free paper   02/3141 YL – 5 4 3 2 1 0

Volume Editors

Dr. Ronit Satchi-Fainaro
Harvard Medical School
and Children's Hospital
Boston Department of Surgery
Vascular Biology Program
1 Blackfan Circle
Boston, MA 02115, USA
ronit.satchi-fainaro@childrens.harvard.edu

Prof. Ruth Duncan
Welsh School of Pharmacy
Cardiff University
Redwood Building
King Edward VII Avenue
Cardiff CF 10 3XF, UK
DuncanR@cf.ac.uk

Editorial Board

Prof. Akihiro Abe
Department of Industrial Chemistry
Tokyo Institute of Polytechnics
1583 Iiyama, Atsugi-shi 243-02, Japan
aabe@chem.t-kougei.ac.jp

Prof. A.-C. Albertsson
Department of Polymer Technology
The Royal Institute of Technology
10044 Stockholm, Sweden
aila@polymer.kth.se

Prof. Ruth Duncan
Welsh School of Pharmacy
Cardiff University
Redwood Building
King Edward VII Avenue
Cardiff CF 10 3XF, UK
DuncanR@cf.ac.uk

Prof. Karel Dušek
Institute of Macromolecular Chemistry,
Czech
Academy of Sciences of the Czech Republic
Heyrovský Sq. 2
16206 Prague 6, Czech Republic
dusek@imc.cas.cz

Prof. W. H. de Jeu
FOM-Institute AMOLF
Kruislaan 407
1098 SJ Amsterdam, The Netherlands
dejeu@amolf.nl
and Dutch Polymer Institute
Eindhoven University of Technology
PO Box 513
5600 MB Eindhoven, The Netherlands

Prof. Jean-François Joanny
Physicochimie Curie
Institut Curie section recherche
26 rue d'Ulm
75248 Paris cedex 05, France
jean-francois.joanny@curie.fr

Prof. Hans-Henning Kausch
Ecole Polytechnique Fédérale de Lausanne
Science de Base
Station 6
1015 Lausanne, Switzerland
kausch.cully@bluewin.ch

Prof. Shiro Kobayashi
R & D Center for Bio-based Materials
Kyoto Institute of Technology
Matsugasaki, Sakyo-ku
Kyoto 606-8585, Japan
kobayash@kit.ac.jp

Prof. Kwang-Sup Lee
Department of Polymer Science & Engineering
Hannam University
133 Ojung-Dong Daejeon,
306-791, Korea
kslee@hannam.ac.kr

Prof. L. Leibler
Matière Molle et Chimie
Ecole Supérieure de Physique
et Chimie Industrielles (ESPCI)
10 rue Vauquelin
75231 Paris Cedex 05, France
ludwik.leibler@espci.fr

Prof. Timothy E. Long
Department of Chemistry
and Research Institute
Virginia Tech
2110 Hahn Hall (0344)
Blacksburg, VA 24061, USA
telong@vt.edu

Prof. Ian Manners
School of Chemistry
University of Bristol
Cantock's Close
BS8 1TS Bristol, UK
Ian.Manners@bristol.ac.uk

Prof. Martin Möller
Deutsches Wollforschungsinstitut
an der RWTH Aachen e.V.
Pauwelsstraße 8
52056 Aachen, Germany
moeller@dwi.rwth-aachen.de

Prof. Oskar Nuyken
Lehrstuhl für Makromolekulare Stoffe
TU München
Lichtenbergstr. 4
85747 Garching, Germany
oskar.nuyken@ch.tum.de

Prof. E. M. Terentjev
Cavendish Laboratory
Madingley Road
Cambridge CB 3 OHE, UK
emt1000@cam.ac.uk

Prof. Brigitte Voit
Institut für Polymerforschung Dresden
Hohe Straße 6
01069 Dresden, Germany
voit@ipfdd.de

Prof. Gerhard Wegner
Max-Planck-Institut
für Polymerforschung
Ackermannweg 10
Postfach 3148
55128 Mainz, Germany
wegner@mpip-mainz.mpg.de

Prof. Ulrich Wiesner
Materials Science & Engineering
Cornell University
329 Bard Hall
Ithaca, NY 14853, USA
ubw1@cornell.edu

Advances in Polymer Science
Also Available Electronically

For all customers who have a standing order to Advances in Polymer Science, we offer the electronic version via SpringerLink free of charge. Please contact your librarian who can receive a password or free access to the full articles by registering at:

springerlink.com

If you do not have a subscription, you can still view the tables of contents of the volumes and the abstract of each article by going to the SpringerLink Homepage, clicking on "Browse by Online Libraries", then "Chemical Sciences", and finally choose Advances in Polymer Science.

You will find information about the

- Editorial Board
- Aims and Scope
- Instructions for Authors
- Sample Contribution

at springeronline.com using the search function.

Contents

Polymer Therapeutics: Polymers as Drugs, Drug and Protein Conjugates and Gene Delivery Systems: Past, Present and Future Opportunities
R. Duncan · H. Ringsdorf · R. Satchi-Fainaro 1

Polymers as Drugs
P. K. Dhal · S. R. Holmes-Farley · C. C. Huval · T. H. Jozefiak 9

Domino Dendrimers
R. J. Amir · D. Shabat 59

PEGylation of Proteins as Tailored Chemistry for Optimized Bioconjugates
G. Pasut · F. M. Veronese 95

Gene Delivery Using Polymer Therapeutics
E. Wagner · J. Kloeckner 135

Author Index Volumes 101–192 175

Subject Index 199

Contents of Volume 193

Polymer Therapeutics II

Volume Editors: Ronit Satchi-Fainaro, Ruth Duncan
ISBN: 3-540-29211-X

Polymer Therapeutics for Cancer:
Current Status and Future Challenges
R. Satchi-Fainaro · R. Duncan · C. M. Barnes

Nanostructured Devices Based on Block Copolymer Assemblies
for Drug Delivery: Designing Structures for Enhanced Drug Function
N. Nishiyama · K. Kataoka

The EPR Effect and Polymeric Drugs:
A Paradigm Shift for Cancer Chemotherapy in the 21st Century
H. Maeda · K. Greish · J. Fang

Molecular-Scale Studies on Biopolymers Using Atomic Force Microscopy
J. S. Ellis · S. Allen · Y. T. A. Chim · C. J. Roberts · S. J. B. Tendler ·
M. C. Davies

Polymer Genomics
A. V. Kabanov · E. V. Batrakova · S. Sherman · V. Y. Alakhov

Polymer Therapeutics: Polymers as Drugs, Drug and Protein Conjugates and Gene Delivery Systems: Past, Present and Future Opportunities

Ruth Duncan[1] (✉) · Helmut Ringsdorf[2] · Ronit Satchi-Fainaro[3]

[1] Centre for Polymer Therapeutics, Welsh School of Pharmacy, Cardiff University, Redwood Building, King Edward VII Avenue, Cardiff CF10 3XF, UK
DuncanR@cf.ac.uk

[2] University of Mainz, Institute of Organic Chemistry, Duesbergweg 10–14, 55099 Mainz, Germany

[3] Children's Hospital Boston and Harvard Medical School, Vascular Biology Program, Department of Surgery, 1 Blackfan Circle, Karp Family Research Laboratories, Floor 12, Boston, Massachusetts 02115, USA

1	Historical Perspective	2
2	Current Status	3
3	Future Opportunities and Challenges	5
	References	6

Abstract As the 21st century begins we are witnessing a paradigm shift in medical practice. Whereas the use of polymers in biomedical materials applications – for example, as prostheses, medical devices, contact lenses, dental materials and pharmaceutical excipients – is long established, polymer-based medicines have only recently entered routine clinical practice [1–4]. Importantly, many of the innovative polymer-based therapeutics once dismissed as interesting but impractical scientific curiosities have now shown that they can satisfy the stringent requirements of industrial development and regulatory authority approval. The latter demand on one hand a cost-effective and profitable medicine or diagnostic, and on the other hand, a safe and efficacious profile that justifies administration to patients.

The first clinical proof of concept with polymer therapeutics has coincided with the explosion of interest in the fashionable area called "nanotechnology". This has resulted in exponential growth in the field, and an increasing number of polymer chemists are turning their attention to the "bio-nano" arena. An attempt to define "nanotechnology" is beyond the scope of this review, but suffice it to say there is widespread agreement that application of nanotechnology to medicine, either via miniaturisation or synthetic polymer and supramolecular chemistry to construct nano-sized assemblies [5, 6], offers a unique opportunity to design improved diagnostics, preventative medicines, and more efficacious treatments of life-threatening and debilitating diseases. It is thus timely for this volume of Advances in Polymer Science to review the field that has been named "polymer therapeutics" (Fig. 1).

The term "polymer therapeutics" [1] has been adopted to encompass several families of constructs all using water-soluble polymers as components for design; polymeric

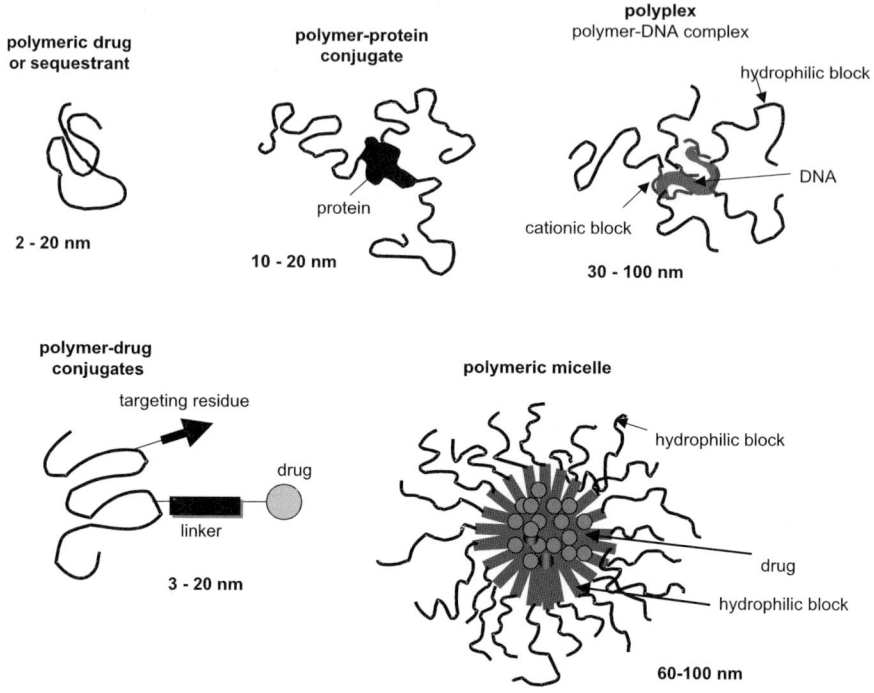

Fig. 1 Schematic showing the families of polymer constructs called "polymer therapeutics"

drugs [3, 7], polymer-drug conjugates [1, 8], polymer-protein conjugates [2, 9], polymeric micelles to which a drug is covalently bound [10], and those multi-component polyplexes being developed as non-viral vectors [11]. From an industrial standpoint, these nanosized medicines are more like new chemical entities than conventional "drug-delivery systems or formulations" which simply entrap, solubilise or control drug release without resorting to chemical conjugation. In this issue of Advances in Polymer Science, the current status of those technologies in preclinical and clinical development is reviewed, together with presentation of an emerging area of novel synthetic chemistry – the new field of polymer genomics – and also a description of some of the sophisticated analytical methods being developed to characterise complex polymer constructs.

1
Historical Perspective

The use of polymers in medicine is not new. Undoubtedly, natural polymers have been used as components of herbal remedies for several millennia. Modern pharmacognosy is currently more carefully identifying specific natural-product macromolecular drugs and beginning to more rigorously define the molecular basis of their mechanisms of action. The notion of synthetic, water-soluble polymers as macromolecular drugs or components of injectible drug delivery systems has, in contrast, a relatively short history – not surprising given the infancy of polymer science itself. The efforts of Hermann Staudinger and his contemporaries led to the birth of polymer science in the 1920s – less than a hundred years ago [12–14]. Moreover, it wasn't until 1953 that Staudinger was honoured with the first "polymer" Nobel Prize "for his discoveries in the field of macromolecular chemistry". Coincidentally, this is the same year that Watson and Crick published their *Nature* articles on the structure of DNA [15]. Around this time we saw the beginning of water-soluble synthetic polymers as healthcare aids for parenteral administration. During the Second World War synthetic polymeric plasma expanders were widely adopted (e.g. poly(vinylpyrolidone)). Before long the first polymer-drug conjugates appeared (e.g. mescaline-*N*-vinylpyrolidine conjugates with drug attached via non-degradable or enzymatically degradable (gly-leu) side chains [16]). Biologically active polymeric drugs also started to gain popularity [17], and divinylether-maleic anhydride copolymer (pyran copolymer) was tested clinically as an anticancer agent in the 1960s. It failed in early clinical trials due to its severe toxicity, and later it was discovered that deleterious effects were related to subtle changes in polymer molecular weight and administration via the intravenous route [18]. Building on the lessons learnt in these early studies, modified polysaccharides, synthetic polypeptides and synthetic polymers have since all been successfully transferred into the market as polymeric drugs. In fact, it was pioneering work that began to emerge in the 1970s that began to lay the foundations for a clearly defined chemical and biological rationale for the design of polymeric drugs, polymer-protein conjugates [9] and polymer-drug conjugates [8, 19, 20].

2
Current Status

Efforts in the 1970s and 1980s allowed rational design (bearing in mind the proposed use and pathophysiology of the disease target) of the first polymer therapeutic candidates that later entered clinical testing. Translation to the clinic solved for the first time many important challenges relating to specific product development of polymertherapeutics: industrial-scale manufacture; development of "validated" analytical techniques required to confirm identity and batch-to-batch reproducibility of these often heterogeneous, hybrid macromolecular constructs; and the development of pharmaceutical formulations able to ensure shelf-life stability and rapid solubilisation of particle-free solutions for safe injection. Definition of preclinical toxicological protocols able to ensure the degree of safety was also needed to justify clinical trials and the optimisation of clinical protocols (dose and frequency of dosing) is still ongoing for many products.

The first poly(ethyleneglycol) (PEG)ylated proteins were approved by regulatory authorities for routine clinical use in the early 1990s (reviewed in this volume by Pasut and Veronese: "Pegylation of Proteins as Tailored Chemistry for Optimized Bioconjugates"): PEG-adenosine deaminase used to treat acute immunodeficiency syndrome and PEG-L-asparaginase to treat acute lymphoblastic leukaemia. At the same time in Japan, a stryene-co-maleic anhydride conjugate of the anticancer protein neocarzinstatin called SMANCS, developed by Maeda and colleagues, was successfully used as a treatment of patients with primary liver cancer (a very difficult disease to treat) and this led to market approval for the treatment of this disease. In this case the aim of polymer conjugation was to hydrophobise the protein, thus allowing dispersion in a phase contrast agent Lipiodol that is used for patient imaging. The formulation is administered locally via the hepatic artery. During his research, Maeda also discovered the passive tumour-targeting phenomenon called the "enhanced permeability and retention effect" (EPR effect). This phenomenon is attributed to two factors: the disorganised pathology of angiogenic tumour vasculature with its discontinuous endothelium leading to hyperpermeability towards circulating macromolecules, and the lack of effective tumour lymphatic drainage, which leads to subsequent macromolecular accumulation. It is now well established that long circulating macromolecules including polymer conjugates, and even polymer-coated liposomes, accumulate passively in solid tumour tissue by the EPR effect after intravenous administration and can increase tumour concentration manyfold (reviewed in this volume by Maeda et al.: "The EPR Effect and Polymeric Drugs: A Paradigm Shift in Cancer Chemotherapy").

Throughout the 1990s a steady stream of polymeric drugs began to emerge (reviewed in this volume by Dhal et al.: "Polymers as Drugs"). These include a number of products including a synthetic random copolymer of L-alanine, L-lysine, L-glutamic acid and L-tyrosine (M_w = 5000–11 000 g/mol) given subcutaneously to treat multiple sclerosis patients and also those poly(allylamine)s developed clinically as polymeric sequestrants for oral administration. In addition, a growing number of compounds have entered clinical trials. They include dextrin-2-sulfate (M_w = 25 000 g/mol) given intraperitoneally to treat HIV-1 in patients, and most recently, the first dendrimer-based drug tested clinically, which is also a vaginal anti HIV virucide.

The first synthetic polymer anticancer drug conjugate entered clinical trials in 1994. This was an N-(2-hydroxypropyl)methacrylamide (HPMA) copolymer conjugate of doxorubicin [21, 22]. Since then, five more HPMA copolymer conjugates have progressed into the clinic, and the first conjugate bearing antiangiogenic therapy is now being tested in vivo [23]. Anticancer conjugates based on other polymeric carriers including poly(glutamic acid), PEG and polysaccharides are also now in clinical trials, and it is anticipated that the first product in this class will appear very soon (reviewed here in Satchi-Fainaro et al.: "Polymer Therapeutics as Anticancer Treatments: Current Status and Future Challenges"). An alternative approach for targeted delivery of anticancer agents utilises block copolymer micelles within which the anticancer drug can be simply entrapped or covalently bound. Of this type there are currently three systems in early clinical trials (reviewed in Nishiyama and Kataoka, "Nanostructured Devices Based on Block Copolymer Assemblies for Drug Delivery: Structural Design for Enhancing Drug Function").

With growing appreciation of the molecular basis of disease in the late 1980s, the hope of "gene therapy" began to gain momentum. While the viral vectors are still preferred for gene delivery, there has been a continuing hope that polymeric non-viral vectors can become a feasible alternative – i.e. biomimetics delivering DNA safely without the threat of toxicity. Pioneering early research used simple polycationic vectors such as poly(L-lysine) and poly(ethyleneimine). Since then a wide variety of complex multicomponent, polymer-based vectors have been designed as gene delivery systems – see Wagner and Kloeckner, "Gene Delivery Using Polymer Therapeutics" and also elsewhere [24]. With still some distance to the first polymeric viral vectors as marketed products, there is still much to do.

3
Future Opportunities and Challenges

It should not be forgotten that it was only the turn of the last century when Paul Ehrlich proposed the first synthetic small molecules as chemotherapy. Introduction of the first biotechnology and polymer-based products over the last two decades has been greeted with the same suspicion that Ehrlich encountered when introducing modern chemotherapy in his day. Nevertheless, at the present time, the core business of the pharmaceutical industry is obviously low-molecular-weight drugs (both natural product extracts and synthetic drugs) and prodrugs, particularly those that are amenable to oral administration providing convenience for the patient.

The fact that macromolecular drugs, such as proteins, polymer therapeutics and genes, are not orally bioavailable, coupled with their chemical complexity and the perceived difficulties in realising them in practice made them unattractive development candidates for many large pharmaceutical companies until the end of the 20th century. Observation that the FDA approved more macromolecular drugs and drug-delivery systems than small molecules as new medicines in 2002/2003 suggests that the tide has now turned.

Now that we are in the 21st century, the time is ripe to build on the lessons learnt over the last few decades, and the increased efforts of polymer chemists working in multidisciplinary teams will surely lead to the design of improved second-generation polymer therapeutics. The polymer community's interest in synthetic and supramolecular chemistry applied to biomedical applications has never been greater. This has in part been due to the rise in interest in using dendrimers and nanotubes for applications in drug delivery (reviewed in this volume by Amir and Shabat:"Domino Dendrimers" and [21]) and not least the need for bioresponsive polymers that can be designed as (3-D) scaffolds for tissue engineering. Innovative polymer synthesis is leading to many new materials, but while they provide exciting opportunities, they also present challenges for careful characterisation of biological and physicochemical characterisation. These two important areas are reviewed in this volume.

For clinical use, it is essential to identify biocompatible synthetic polymers that will not be harmful in relation to their route, dose and frequency of administration. For many years, the general cytotoxicity, haematotoxicity and immunogenicity (cellular and humoral) of water-soluble polymers has been widely studied. Before clinical studies, rigorous preclinical toxicity testing of the candidate has also been mandatory. However, it is becoming evident that synthetic polymers can display many subtle and selective effects on cells affecting a diverse range of biochemical processes. These effects may be relatively weak so they do not result in major toxicity. Studies have recently commenced that assess the pharmacogenomic effects of polymers, and this important, emerging field is reviewed here by Kabanov et al. ("Polymer Genomics"). Development of analytical techniques able to accurately characterise polymer therapeutics in terms of identity, strength, stability and structure in real time (to allow correlation with biological properties) has proved a real challenge in itself. However, atomic-force microscopy has already begun to demonstrate the ability to provide structural and physicochemical information for a wide range of synthetic and bio-polymers. The latest developments in the latter area are described here by Davies "Characterisation of polymer constructs by Real Time Molecular AFM investigations".

This volume highlights some of the key areas of research and development relating to synthesis, characterisation and use of polymer therapeutics. For those new to the field, the text should be read in parallel with the historical milestone publications (see the bibliography), including papers published in Advances in Polymer Science (for example [25, 26]) and elsewhere [8, 19]. There are also several recent reviews that are essential reading for the expert and newcomer alike [27, 28].

References

1. Duncan R (2003) The dawning era of polymer therapeutics. Nat Rev Drug Discov 2:347–360
2. Harris JM, Chess RB (2003) Effect of pegylation on pharmaceuticals. Nat Rev Drug Discov 2:214–221
3. Dhal PK, Holmes-Farley SR, Mandeville WH, Neenan TX (2002) Polymeric drugs. In: Encyclopedia of Polymer Science and Technology, 3rd edn. Wiley, New York, pp 555–580

4. Duncan R (2003) Polymer-drug conjugates. In: Budman D, Calvert H, Rowinsky E (eds) Handbook of Anticancer Drug Development. Lippincott, Williams & Wilkins, Philadelphia, pp 239–260
5. Editorial. Nanomedicine: grounds for optimism. Lancet, pp 362, 673; (2004) NIH Roadmap for Nanomedicines, Willis RC (2004) Good things in small packages. Nanotech advances are producing mega-results in drug delivery. Modern Drug Discov p 30–36. European Science Foundation Policy Briefing (2005) ESF Scientific Forward Look on Nanomedicine 23 February 2005 (www.esf.org)
6. Ferrari M (2005) Cancer nanotechnology: opportunities and challenges. Nat Rev Cancer 5:161–171
7. Gebelein CG, Carraher CE (1985) Bioactive polymeric systems. Plenum, New York
8. Ringsdorf H (1975) Structure and properties of pharmacologically active polymers. Polym J Sci Polymer Symp 51:135–153
9. Davis FF (2002) The origin of pegnology. Adv Drug Del Rev 54:457–458
10. Kakizawa Y, Kataoka K (2002) Block copolymer micelles for delivery of gene and related compounds. Adv Drug Deliv Rev 54:203
11. Wagner E (2004) Strategies to improve DNA polyplexes for in vivo gene transfer: will "artificial viruses" be the answer? Pharm Res 21:8–14
12. Ringsdorf H (2004) Hermann Staudinger and the future of polymer research: jubilees – beloved occasions for cultural piety. Angew Chem Int Ed 43:1064–1076
13. Morawetz H (1985) Polymers: the origins and the growth of a science. Wiley, New York
14. Lehn JM (1995) Supramolecular chemistry: concepts and perspectives. Wiley, New York
15. Watson JD, Crick FH (1953) Molecular structure of nucleic acids; a structure for deoxyribose nucleic acid. Nature 171:737–8
16. Jatzkewitz H (1955) Peptamin (glycyl-L-leucyl-mescaline) bound to blood plasma expander (polyvinylpyrrolidone) as a new depot form of a biologically active primary amine (mescaline). Naturforsch Z 10b:27–31
17. Breslow DS (1976) Biologically active synthetic polymers. Pure Appl Chem 46:103–13 Seymour LW (1991) Synthetic polymers with intrinsic anticancer activity. J Bioact Comp Polymers 6:178–216
18. Regelson W (1986) Advances in intraperitoneal (intracavitary) administration of synthetic polymers for immunotherapy and chemotherapy. J Bioact Compat Polymers 1:84–106
19. Gros L, Ringsdorf H, Schupp H (1981) Polymeric antitumour agents on a molecular and cellular level. Angew Chem Int Ed 20:301–323
20. de Duve C, de Barsy T, Poole B, Trouet A, Tulkens P, van-Hoof F (1974) Lysosomotropic agents. Biochem Pharmacol 23:2495–2531
21. Duncan R (2003) Polymer-drug conjugates. In: Budman D, Calvert H, Rowinsky E (eds) Book of anticancer drug development. Lippincott, Williams & Wilkins, Philadelphia, pp 239–260
22. Duncan R (2005) N-(2-Hydroxypropyl)methacrylamide copolymer conjugates. In: Kwon GS (ed) Polymeric drug delivery systems. Dekker, New York, pp 1–92
23. Satchi-Fainaro R, Puder M, Davies JW, Tran HT, Sampson DA, Greene AK, Corfas G, Folkman J (2004) Targeting angiogenesis with a conjugate of HPMA copolymer and TNP-470. Nature Med 10:255–261
24. Pack DW, Hoffman AS, Pun S, Stayton PS (2005) Design and development of polymers for gene delivery. Nature Rev Drug Discov 4:581–593
25. Duncan R, Kopecek J (1984) Soluble synthetic polymers as potential drug carriers. Adv Polymer Sci 57:51–101

26. Bader H, Dorn K, Hupfer B, Ringsdorf H (1985) Polymeric monolayers and liposomes as models for biomembranes. How to bridge the gap between polymer science and membrane biology? Adv Polymer Sci 64:1–62
27. Torchilin VP (2005) Recent advances with liposomes as pharmaceutical carriers. Nature Rev Drug Discov 4:145–160
28. Duncan R, Izzo L (2005) Dendrimer biocompatibility and toxicity. In: Florence AT (ed) Advanced drug delivery reviews special issue on dendrimers. in press

Polymers as Drugs

Pradeep K. Dhal (✉) · S. Randall Holmes-Farley · Chad C. Huval · Thomas H. Jozefiak

Drug Discovery and Development, Genzyme Corporation, 153 Second Avenue, Waltham, MA 02451, USA
Pradeep.dhal@genzyme.com

1	Introduction	11
1.1	Pharmaceutically Active Polymers	12
2	**Polymers for Molecular Sequestration**	13
2.1	Polymeric Drugs for the Sequestration of Inorganic Ions	14
2.1.1	Polymeric Sequestrants for Potassium Ions: A Treatment for Hyperkalemia	14
2.1.2	Sequestration of Phosphate Ions: Polymeric Drugs for Chronic Renal Failure	15
2.1.3	Sequestration of Iron: Polymeric Drugs for the Treatment of Iron Overload Disorders	21
2.2	Bile Acid Sequestrants: Polymeric Cholesterol-Lowering Drugs	25
3	**Polyvalent Interactions and Anti-Infective Polymeric Drugs**	30
3.1	Polymeric Sequestrants of Toxins	32
3.1.1	Sequestration of *Clostridium Difficile* Toxin	32
3.1.2	Sequestration of Anthrax Toxin	36
3.2	Polyvalent Ligands as Antiviral Agents	37
3.2.1	Inhibition of Influenza Virus	37
3.2.2	Anionic Polymers as Anti-HIV Agents	40
3.3	Polyvalent Antimicrobial Agents	40
4	**Polymeric Drugs for the Treatment of Autoimmune Diseases**	44
4.1	Polymeric Drugs to Treat Multiple Sclerosis	44
5	**Polymeric Anti-Obesity Drugs**	47
5.1	Obesity and Medical Need	47
5.2	Inhibition of Lipase to Control Digestion and Absorption of Dietary Fat	48
5.3	Polymeric Fat Binder and Dual Acting Polymeric Lipase Inhibitor-Fat Binder	49
6	**Polymer Therapy for Sickle Cell Disease**	52
6.1	Sickle Cell Disease	52
6.2	Non-Ionic Surfactant for Treating Sickle Cell Disease	52
7	**Conclusions and Outlook**	53
	References	54

Abstract Polymeric drugs are defined as polymers that are active pharmaceutical ingredients, i.e., they are neither drug carriers nor prodrugs. In general, the underlying concept behind these therapeutic agents is the utilization of high molecular weight and functional characteristics of polymers to selectively recognize, sequester, and remove low molecular weight and macromolecular disease causing species in the intestinal fluid. The high molecular weight nature of these therapeutically relevant polymers makes them systemically non-absorbed, thus providing a number of advantages including long-term safety profiles over traditional small molecule drug products. Furthermore, multiple functional groups in the polymers incorporate polyvalent binding interactions that can result in pharmaceutical properties not found in small molecule drugs. This article summarizes some of the most recent efforts for the discovery and development of polymeric drugs that have proceeded from discovery phase to market place. Examples include sequestration of low molecular weight species such as bile acids, phosphate, and iron ions as well as polyvalent interactions to bind toxins, viruses, and bacteria as well as polymeric enzyme inhibitors and fat binders as anti-obesity agents. Furthermore, use of functional polymers to treat autoimmune disease and sickle cell anemia has been reviewed.

Keywords Bioactive polymers · Drugs · Non-absorbed · Polyvalent · Sequestrants

Abbreviations

GI	gastrointestinal
SAR	structure-activity relationship
LDLc	low-density lipoprotein cholesterol
HMG-CoA	3-hydroxy-3-methyl glutaryl coenzyme A
BAS	bile acid sequestrant
C. difficile	Clostridium difficile
PA	protective antigen
EF	edema factor
LF	lethal factor
Tyr	tyrosine
Trp	tryptophan
VAP	viral attachment proteins
HA	hemagglutinin
SA	sialic acids
Kd	dissociation constant
M	molar
mM	milli molar
pM	pico molar
HIV	human immune virus
MRSA	methicillin-resistant Staphylococcus aureus
C. parvus	Cryptosporadium parvum
VRE	vancomycin-resistant Enterococi
ROMP	ring opening metathesis polymerization
S. aureus	Staphylococcus aureus
RA	rheumatoid arthritis
MS	multiple sclerosis
TNF-α	tumor necrosis factor alpha
GA	glatiramer acetate

MBP	myelin basic protein
CNS	central nervous system
FDA	Food and Drug Administration
EAE	experimental allergic Encephalomyelitis
TAG	triacyl glycerides
PEO	polyethylene oxide
PPO	polypropylene oxide

1
Introduction

During the last three decades the role of polymers in biomedicine has seen significant growth. The unique physico-chemical properties offered by polymeric materials have been exploited in a variety of biomedical applications. A wide range of functional polymers with novel structural architectures have been synthesized during this period and have been evaluated in biological environments [1, 2]. The use of polymers materials in drug delivery and as components in artificial organs, tissue engineering, medical devices, and dentistry is well known [3]. However, an increasingly important aspect of the field of biomedical polymers is the recognition of the role of polymers as new and novel chemical entities for therapeutic application. For this purpose, the polymer may be intrinsically bioactive, or can be utilized as a carrier for site specific and sustained delivery of chemo- and biotherapeutic agents [4].

Following the original work of Ringsdorf presenting the concept of site-targeted polymeric drugs, a large body of scientific literature has appeared, affirming the role of functional polymers as vehicles for therapeutic agents against a variety of diseases. These polymer-drug conjugate systems have enabled the delivery of small molecule drugs and biotherapeutic agents at a controlled rate, and have achieved the targeted delivery of chemotherapeutic agents to specific sites (e.g. to minimize dose-dependent toxicity and enhance selectivity for anti-neoplastic agents) [5–7]. More recently, several functional polymers have been evaluated as non-viral vectors for the delivery of genetic materials for gene therapy applications [8, 9]. The research efforts in the area of polymer-based drug delivery systems, including polymer-drug conjugates, has brought about considerable progress in this area including clinically approved products. Numerous high-quality research papers and excellent review articles dealing with this aspect of biomedical polymers have been published over the last two decades [10–12]. Since several books and review articles pertaining to polymeric drug *delivery* systems and polymer-drug conjugates have been published, this article has been limited to the review of biomedical polymers that act as active pharmaceutical ingredients.

1.1
Pharmaceutically Active Polymers

While functional polymers as carriers for therapeutic agents have been extensively studied, examples of polymers acting as active pharmaceutical ingredients are relatively scant. Early studies pertaining to the use of polymers as therapeutic agents include the use of poly(ethylenesulfonate) as a topical anticoagulant and as an antitumor agent [13, 14]. Although this polymer had shown anti-tumor activity against a broad selection of tumor cell lines, it lacked an acceptable therapeutic index. Almost three decades ago a relatively simple class of polymers, maleic anhydride-divinyl ether copolymer, was studied for their effect on tumor cell lines. It was demonstrated that these anionic polymers are able to modulate the immune system by stimulating T-cell activity. Although the therapeutic effects of anionic polymers were found to be modest, they certainly foretold the interesting adaptations of polymers to such complicated disease states as cancer [15, 16]. In spite of these early promises, effective research effort to develop intrinsically bioactive polymers as therapeutic agents is a relatively recent phenomenon [17, 18]. As potential therapeutic agents, the high molecular weight characteristics of polymers would appear to offer several advantages over classical small molecule drug candidates. Although polymers do not fit most "drug-like" definitions and inherently violate "Lipinski's Rules", a re-evaluation of the attributes of polymers highlights many possible benefits of polymeric pharmaceuticals that are unattainable with traditional small molecule drugs. These benefits may include: lower toxicity, greater specificity of action, and enhanced activity due to multiple interactions with disease targets (polyvalency). In spite of these potential benefits, the concept of polymeric drugs has been a subject of considerable skepticism among drug discovery and development scientists. As a class of potential pharmacophores for drug development, synthetic polymers have been thought to be uninteresting by medicinal chemists and regulatory authorities. Some of the underlying concerns attributed to polymers as new chemical entities for therapeutic applications include the issue of broad molecular weight distribution (polydispersity) and compositional and structural (microstructure including stereochemistry) heterogeneity. These shortcomings were considered to impede drug development and regulatory approval. Furthermore, the high molecular weight characteristics of polymers would potentially undermine their systemic absorption through oral administration (oral bioavailability). However, these seemingly detrimental pharmacological features of polymers can indeed be exploited to design and develop novel therapeutic agents for disease conditions where low molecular weight drugs have either failed or exhibited inadequate therapeutic benefits [19].

The systemic non-absorption characteristics of high molecular weight polymers taken through the oral route may offer therapeutic benefits where it is desirable to minimize (even stop) systemic exposure of drug substances. For example, by combining this non-systemic bioavailability characteristic with the ability of macromolecular sorbents to selectively recognize and sequester molecular and macromolecular components in gastrointestinal fluids, it has become possible to develop a powerful new class of therapeutic agents that can selectively bind and remove detrimental species (attributed to several disease indications) from the gastrointestinal (GI) tract. As with any pharmaceutical agents, sequestration of each target molecule or pathogen requires a unique strategy and this strategy depends not only on the chemical nature of the target, but the location, concentration, and quantity of the target to be removed at a therapeutically acceptable dose.

Some of the most recent efforts in the area of discovery and development of polymers as therapeutic agents are reviewed in the present article. The case studies presented in this article concentrate on the development of polymeric drugs for the sequestration of low molecular weight species such as bile acids, phosphate, and iron as well as polyvalent interactions to bind toxins, viruses, and bacteria. Furthermore, the use of polymers as immunomodulating agents to treat autoimmune diseases by inhibiting specific biological antigen binding events is highlighted. Recent developments in these areas of polymeric drugs including examples of approved and marketed pharmaceutical products are presented.

2
Polymers for Molecular Sequestration

A number of potentially detrimental substances implicated for various disease states, are present or circulate in the GI tract. These species can either enter the body with food, or from the environment (exogenic), or they may be produced in the human body as a result of metabolism (endogenic). Effective removal of these detrimental species from the GI tract in a selective manner offers a promising approach to treat a number of diseases. Thus, biologically appropriate and non-toxic polymeric sorbents represent an ideal class of therapeutic agents for this purpose. The non-absorption of these polymeric resins through the intestinal wall would deter systemic exposure and should result in minimal toxicity. By incorporating appropriate functional groups and by modulating the physicochemical characteristics of polymers, a wide range of possibilities to tailor-make polymers with high selectivity and capacity for targeted species are conceivable. In the following sections we provide an overview of a number of polymeric sequestrants that have been discovered and developed during the last few years as therapeutic agents to treat human diseases.

2.1
Polymeric Drugs for the Sequestration of Inorganic Ions

The process of electrolyte homeostasis is the key to critical physiological functions such as myocardial and neurological functions, fluid balance, oxygen delivery, and acid-base balance. Perturbation of this delicate electrolyte balance by either excessive ingestion or impairment of the elimination process due to dysfunctional metabolism can bring about detrimental pathological consequences. Although certain non-renal tissues like muscle and liver contribute to maintaining the electrolyte balance, the kidney is the primary organ responsible for maintaining electrolyte homeostasis [20]. As a result, impairment of renal function is the main factor responsible for electrolyte imbalance and can result in life-threatening metabolic disorders. Thus, the use of non-absorbed polymeric sequestrants to bind these excessive ions (implicated in various pathologic conditions) selectively in the GI tract is, in principle, a novel approach to treat or prevent disease conditions associated with electrolyte imbalance.

2.1.1
Polymeric Sequestrants for Potassium Ions: A Treatment for Hyperkalemia

Potassium ions play a critical role in regulating the transmembrane potential for cellular functions. Therefore, maintenance of the critical ratio of potassium ions between intra- and extracellular fluids is important to all living cells [21]. Hyperkalemia (elevated level of serum potassium, usually greater than 5.0 mEq/L) can result from burn and crush muscle injuries, acidosis, or through the use of anti-hypertension drugs like angiotensin-converting enzyme inhibitors. A rise in serum potassium can manifest moderate to serious health problems such as paresthesias, areflexia, respiratory failure, and bradycardia [22]. Since the kidney is the primary route to remove excessive body potassium, patients with impaired renal function are incapable of maintaining potassium homeostasis. Traditional approaches to treat hyperkalemia include the use of insulin, glucose, sodium bicarbonate, and calcium chloride. However, these treatments have their own shortcomings. For example, excess intake of calcium leads to hypercalcemia, which in turn leads to other complications like myocardial infarction and kidney stones.

Use of an insoluble anionically charged polymer resin to sequester excess potassium ions in the GI tract and their subsequent elimination in the feces is a simple approach that is being used to treat this disease. This approach enables patients to remove excess potassium from their bodies in spite of impaired kidney function. A number of low molecular weight ligands that complex potassium ions are known in the literature [23]. Utilizing this principle of potassium ion binding by organic ligands, a cation-exchange resin based on sodium polystyrene sulfonate (Scheme 1) was developed to

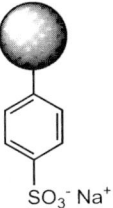

Scheme 1

sequester potassium ion in the GI tract. This polymer, marketed under the brand name Kayexalate has been approved in the United States for the treatment of hyperkalemia since 1975 [24]. However, a potential problem associated with the use of Kayexalate is induction of hypernatremia (elevated serum sodium), since the polymer resin exchanges 1.0 equivalent of potassium for 1.5 equivalents of sodium. Furthermore, cases of intestinal necrosis have been attributed to Kayexalate [25]. The adverse effects associated with this polymeric potassium sequestrant provide an opportunity to develop new generations of potassium sequestering polymers. Unfortunately, no new efforts have been made to develop second generation sequestration therapies to treat hyperkalemia.

2.1.2
Sequestration of Phosphate Ions: Polymeric Drugs for Chronic Renal Failure

The management and control of elevated levels of serum phosphate (hyperphosphatemia) is a critical health concern for patients suffering from chronic and end-stage renal disease [26, 27]. Consequences of inadequate control of serum phosphate level can manifest a number of pathologies of clinical significance. These include soft tissue calcification (leading to cardiac calcification and cardiac-related complication), renal bone disease leading to reduced bone density, and secondary hyperparathyroidism. These detrimental factors make hyperphosphatemia a major risk factor for mortality among patients suffering from end-stage renal disease (e.g., dialysis patients).

The kidney is the primary route for the excretion of phosphate from a healthy human body. Therefore, patients with impaired renal function accumulate phosphate as a result of an imbalance between ingestion and excretion of dietary phosphate. Phosphate binder therapy has been the mainstay for the treatment of hyperphosphatemia. The traditional phosphate binders have been calcium- and aluminum-based agents. These inorganic cations remove phosphate through the formation of corresponding insoluble phosphate salts in the GI tract [28, 29]. Since aluminum and calcium salts have the propensity for systemic absorption through the intestinal mucosal layer, prolonged treatment involving these agents carry the liability of bringing about undesirable toxic and metabolic side effects (such as neurological disorders, cardiac calcifications, etc.) in renal-compromised patients. These shortcomings of calcium- and aluminum-based binders limit their long-term use as phosphate sequestrants.

Non-absorbed cationic polymers as sequestrants for phosphate ions offer an effective and safe approach to treat hyperphosphatemia in renal failure patients. The binding of phosphate ions by polycationic species is a well-studied phenomenon in molecular recognition and supramolecular chemistry research [30]. Towards this end, a great deal of research efforts have been made over the past several years to design and synthesize novel compounds such as oligomeric/macrocyclic amines, ammonium salts, and guanidinium compounds (e.g., Schemes 2-4) as synthetic receptors for phosphate and related anionic guests (*viz.* phosphate, pyrophosphate, and phosphonate anions) [31-33]. Electrostatic interaction is the primary driving force for complexation of phosphate-based anions with these organic cationic hosts. Hydrogen bonding is considered to lend additional binding strength [34]. The underlying principle of the physical organic chemistry of this anion recognition process was recruited to discover non-absorbed, cationic, polymeric hydrogels (such as polymeric amines and guanidinium compounds) that show affinity towards phosphate ions derived from dietary sources. Being non-absorbed, these polymeric sequestrants are confined to the GI tract. Therefore, they act as effective therapeutic agents (free from the side effects associated with calcium and related metal salt-based phosphate binders) to treat hyperphosphatemia.

Scheme 2

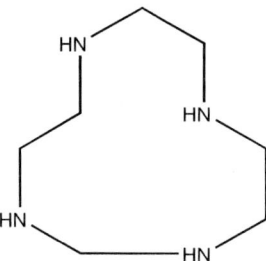

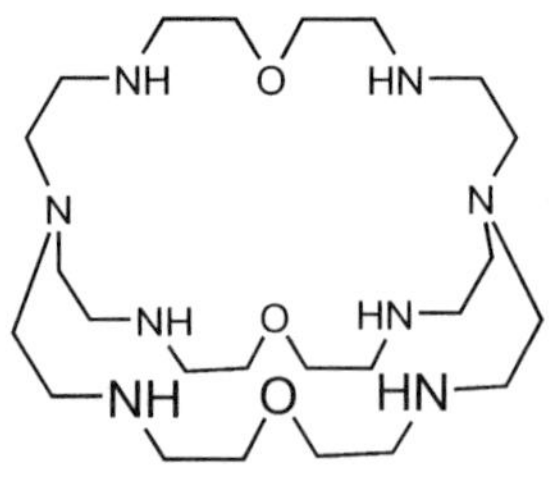

Scheme 3

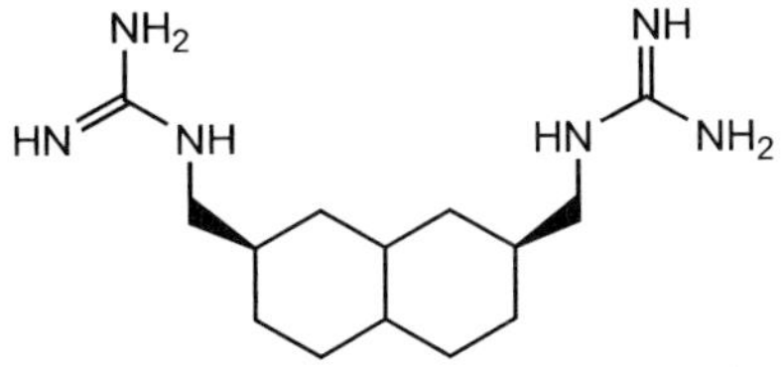

Scheme 4

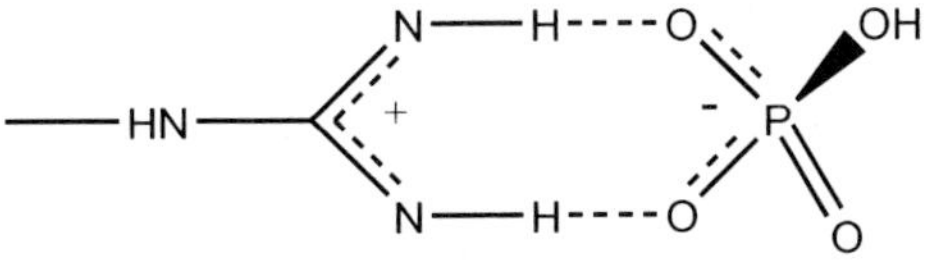

Scheme 5

Binding of phosphate ions to guanidinium groups of arginine residues of proteins (involving two electrostatic bonds and two stereochemically favorable hydrogen bonds, structure Scheme 5) is a well-known and important phenomenon in biological systems [35]. This biological principle of phosphate recognition was utilized by Hider and Rodriguez in designing and synthesizing a series of insoluble polymers containing guanidinium groups as sequestrants for phosphate anions [36]. The synthetic procedure employed to prepare these guanidinium-functionalized polymer resins is illustrated in Fig. 1. Phosphate binding studies involving these polymeric guanidinium salts under in vitro conditions has shown that these polymers bind phosphate selectively in the presence of other competing biologically important anions such as chloride, bicarbonate, etc.

A series of amine containing, polymeric phosphate sequestrants were discovered and systematically investigated in our laboratories. This work finally led to a marketed drug for the treatment of hyperphosphatemia (vide infra). Using the knowledge of phosphate binding properties of macrocyclic polyamines and related compounds as anion receptors (described above), we synthesized a series of amine containing functional polymers. These amine-functionalized polymeric hydrogels were prepared either by postpolymerization crosslinking of polymeric amines, or by the crosslinking copolymerization of appropriate amine containing vinyl monomers [37–39]. By these processes a variety of polymer structures bearing primary, secondary, tertiary amine groups, as well as quaternary ammonium groups were obtained. A general procedure for the preparation of polymeric amine-based hydrogels by the crosslinking of soluble polymers is illustrated in Fig. 2. The main structural repeat units of a selection of representative polymeric amines used in our investigation are summarized in Table 1. The in vitro phosphate binding properties of some of these polymeric amine-based hydrogels are given in Table 2. A systematic structure-activity relationship (SAR) study en-

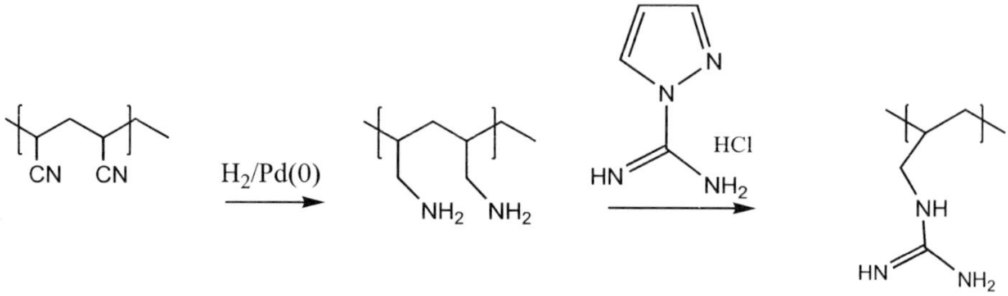

Fig. 1 Synthesis of polymeric guanidinium salts through chemical modification of poly(acrylonitrile) resin

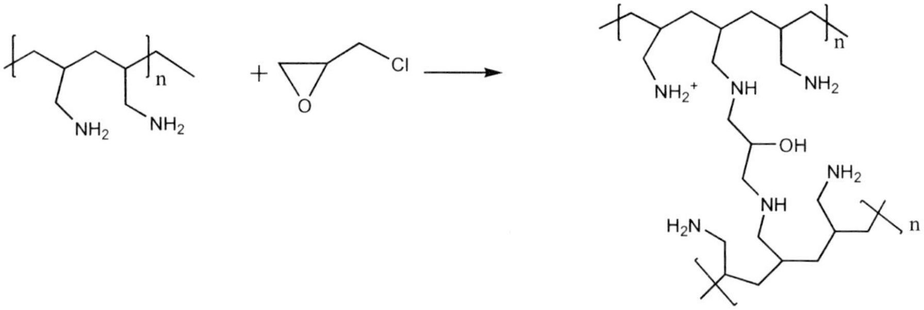

Fig. 2 The general synthetic procedure to prepare amine-functionalized hydrogels by a post-polymerization crosslinking reaction

abled us to identify epichlorohydrin crosslinked polyallylamine (Scheme 6) as the lead, which was advanced to preclinical and clinical development. These polymers are believed to bind phosphate ions through electrostatic and possibly through hydrogen bonding interactions (See Fig. 3).

Like low molecular weight phosphate receptors, the binding strengths and capacities of these polymeric phosphate sequestrants for phosphate ions have been found to depend on the pH of the medium. Since the complexation process involves electrostatic interaction between the polymer-bound ammonium groups and phosphate anions, the optimum level of protonation of amine groups along the polymer chain is key to achieving maximum binding capacity. Due to higher local concentrations of amine groups in polymeric systems, and the divalent (and possibly trivalent) nature of phosphate anions, polymeric amines also exhibit stronger affinity towards phosphate compared to their corresponding small molecule receptor analogs. This enhanced binding affinity of polymers towards phosphate anions can be attributed to a chelating effect. This phenomenon, however, has not been examined in detail. A carefully study of the pH effect may shed further light on the phosphate binding phenomenon in these polymeric systems. Furthermore, since the degree of protonation of polymeric amines is dependent on polymer ar-

Table 1 Structural repeat units of representative polymeric amine precursors used to prepare phosphate sequestrants

Table 2 In vitro equilibrium phosphate-binding properties of amine-functional hydrogels at 5 mM Phosphate

Nature of Polymeric amine	Phosphate bound to polymer (meqv/g)
Polyallylamine/epichlorohydrin	3
Polyethyleneimine/acryloyl chloride	1.2
Diethylenetriamine/epichlorohydrin	1.5
Poly(dimethylaminoethylacrylamide)	0.8
Poly(4-trimethylammoniummethyl)styrene chloride	0.7

chitectures (due to the charge-charge repulsion effect that is prevalent in polyelectrolyte systems), incorporation of appropriate spacing between the amine groups along the polymer chain is another important factor to maximize the concentration of cationic groups, which would in turn influence the phosphate binding capacities and strengths of these polymeric sequestrants [40]. Finally, polymers containing primary amine groups were found to be better phosphate sequestrants than polymers bearing secondary and ter-

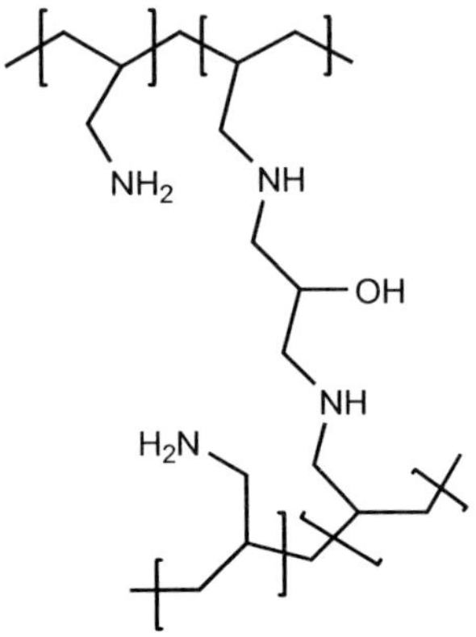

Scheme 6

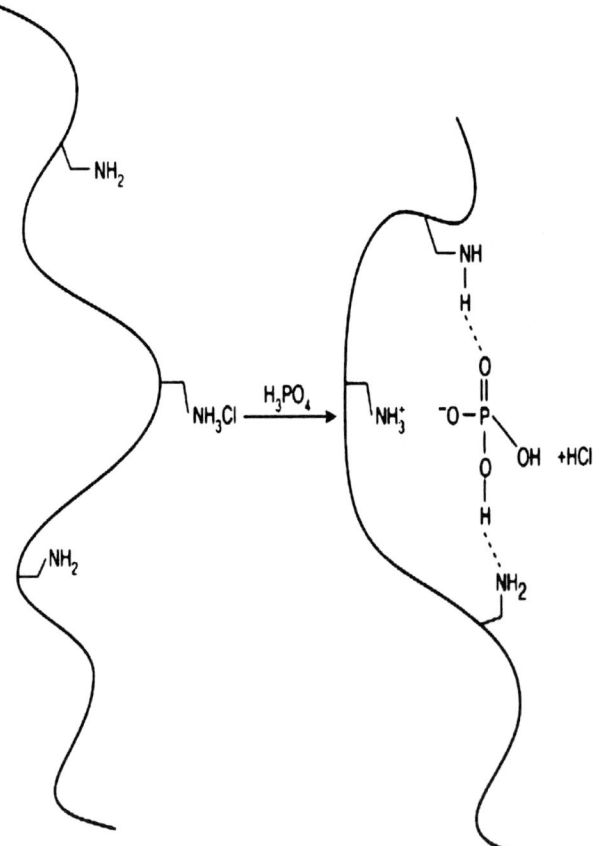

Fig. 3 Binding interactions between polymeric amine hydrochloride gels and phosphate anions

tiary amines, while polymers containing quaternary amines were found to be the poorest sequestrants of phosphate anions.

As mentioned above, SAR studies revealed that crosslinked polyallylamine gels are very potent phosphate-binding polymers and possess properties suitable for pharmaceutical applications. This class of polymers exhibited maximum phosphate binding in the pH range encountered in the milieu of the GI tract. The polymer gels have been found to be non-toxic and are essentially non-absorbed. These polymer gels were very well tolerated in multiple clinical trials with end-stage renal failure patients undergoing hemodialysis [41, 42].

After successful multiphase clinical trials as the first metal-free phosphate sequestrant for the treatment of hyperphosphatemia, polymer Scheme 6 was approved in the United States by the FDA in 1998 under the generic name of sevelamer hydrochloride. It has been marketed since under the brand name Renagel by Genzyme Corporation.

In subsequent years, sevelamer hydrochloride has been approved in Europe, Japan, and a number of other countries. Since its regulatory approval, Renagel has demonstrated effective, long-term control of serum phosphate levels and has shown several clinical benefits over and above traditional inorganic binders for the management of hyperphosphatemia in renal failure patients [43–45]. The ability for improved control of serum phosphate without increasing the exposure to toxic metal ions like aluminum and eliminating the intake of additional calcium offers a number of clinically relevant benefits. For example, without increasing the calcium load or promoting arterial calcification, Renagel may help prevent cardio-vascular complications in patients suffering from end-stage renal disease. Furthermore, Renagel has been found to reduce serum parathyroid hormone and also reduce total and low-density lipoprotein cholesterol (LDLc) in hemodialysis patients. Since cardiovascular events are the most common causes of mortality among dialysis patients, Renagel thus offers a very effective treatment in the management of renal failure [46]. Renagel, represents one of the first tailor-made, polymeric drugs that exhibits prophylactic and therapeutic properties through the selective sequestration and removal of unwanted dietary components in the GI tract without presenting any systemic side effects. These clinically proven benefits of Renagel provide an opportunity to impact patient survival and morbidity as well as to reduce overall health care costs associated with the treatment of end-stage renal disease.

2.1.3
Sequestration of Iron:
Polymeric Drugs for the Treatment of Iron Overload Disorders

Iron is an important metal ion in biological systems. While it is essential for the proper functioning of all living mammalian cells, the presence of excess iron in the body leads to toxic effects [47]. Through the well-known Fenton reaction (Fig. 4) excess iron catalyzes the transformation of molecular oxygen

$$2O_2^{\cdot-} + 2H^+ = H_2O_2 + O_2 \text{ (SOD catalyzed)}$$

$$H_2O_2 + H^+ + Fe^{+2} = HO^{\cdot} + Fe^{+3} + H_2O$$

$$O_2^{\cdot-} + Fe^{+3} = O_2 + Fe^{+2}$$

Fig. 4 Production of reactive oxygen species through an iron-catalyzed Fenton reaction

to oxygen-derived free radicals such as hydroxyl radicals. These oxygen-derived radicals in turn cause damage to many vital biological molecules. This peroxidative tissue or organ damage forms the basis for several pathological conditions including neurodegenerative diseases [48, 49]. Under normal physiological conditions, iron metabolism is tightly conserved with the majority of the iron being recycled within the body. Normal physiology does not provide a mechanism for iron loss, i.e., iron is not normally excreted in urine, feces, or bile. Some loss of iron from the body occurs only through bleeding or normal sloughing of epithelial cells. Certain genetic disorders can, however, lead to increased absorption of dietary iron (as in the case of hemochromatosis) or transfusion-induced iron over load (as in the case of β-thalassaemia and sickle cell anemia) [50, 51].

In the case of hemochromatosis, excess iron can be removed from patients' bodies by venesection. On the other hand, removal of iron using an iron chelator is the only effective way to relieve iron overload in patients with β-thalasemia or sickle cell anemia [52, 53]. Neither of these therapies are optimum for the treatment of these diseases. The current standard of care for iron chelation therapy is desferrioxamine (Scheme 7), which is the only approved iron chelator for the treatment of iron overload conditions in the USA. Desferrioxamine has reportedly been associated with several drawbacks. It has a narrow therapeutic window and due to lack of oral bioavailability, it requires administration for 8–12 h per day by parenteral infusions [54]. Thus, there is a clear need for the discovery and development of a new generation of orally active iron-chelating agents for the treatment of iron overload conditions.

Non-absorbed polymeric ligands that selectively sequester and remove dietary iron from the GI tract would offer an attractive method for the

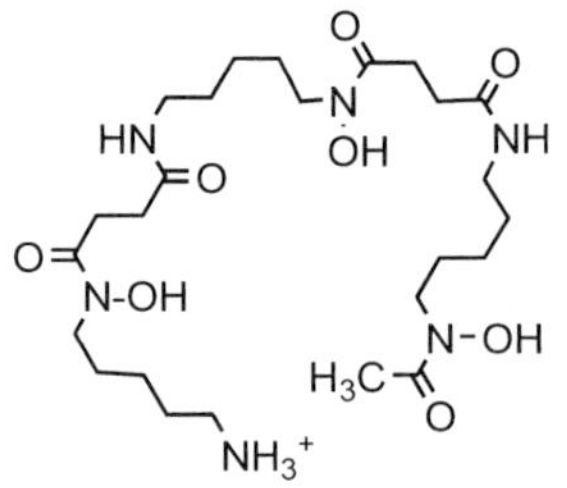

Scheme 7

treatment of certain iron overload conditions. Clinically useful polymeric iron chelators require several features. Because of the importance of other metal ions for normal human physiology, the polymeric ligand must possess high affinity, capacity, and selectivity towards iron. Furthermore, the chelator should be biocompatible, and should not be absorbed from the GI tract. This is particularly important for hemochromatosis. However, binding dietary iron in the GI tract alone may not be sufficient to treat transfusion-related iron overload such as beta thalassemia. For clinical applications, the important properties of chelators include metal ion selectivity and high stability constant for the ligand-metal complex. Since iron exists in two oxidation states [Ferrous (+2) and Ferric (+3)], chelators could be designed to sequester both forms of iron. Thus, the design of pharmaceutically relevant polymeric iron chelators has been based on the knowledge of low molecular weight chelators.

For Fe(II), soft donor atoms (e.g., nitrogen-containing ligands such as bipyridine, Scheme 8 and phenanthroline, Scheme 9) can be employed. Although these ligands are selective for Fe (II), they also possess affinity towards other biologically important divalent metal ions such as Zn(II) and Cu(II). On the other hand, oxyanions like hydroxamates and catecholates are selective towards Fe(III) and these ligands in general show higher selectivity towards trivalent metal ions over divalent metal ions. Nature offers a precedent: natural iron chelators like siderophores such as desferrioxamine (Scheme 7) and enterobactin (Scheme 10) contain hydroxamate and catechol groups respectively, and they exhibit selectivity towards Fe(III) [55, 56].

On the basis of the above criteria, crosslinked polymeric hydrogels containing hydroxamic acid and catechol moieties (Schemes 11 and 12) as well as crosslinked polymeric amines were prepared and were evaluated as iron chelators [57]. Under in vitro conditions, all of these polymers sequester iron at high pH. At lower pH, the polymers containing hydroxamic acids maintained their iron binding properties, while other polymers showed poor iron binding properties. The iron binding isotherms for a hydroxamic acid containing hydrogel at different pH values are shown in Fig. 5. In vivo studies

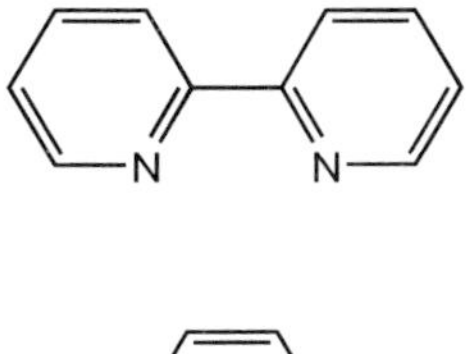

Scheme 8

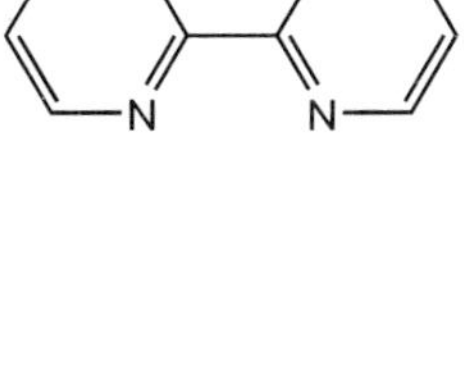

Scheme 9

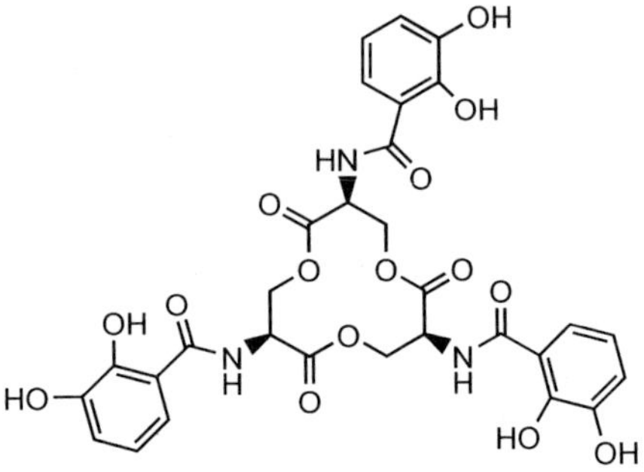

Scheme 10

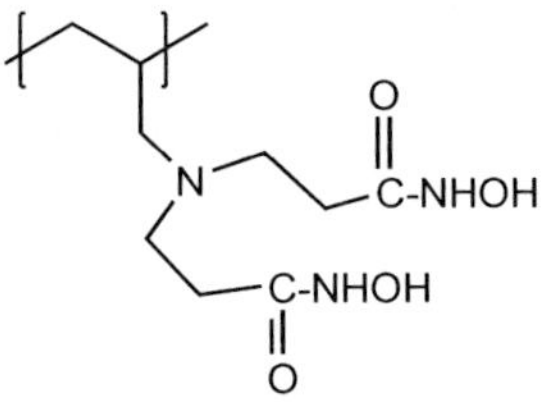

Scheme 11

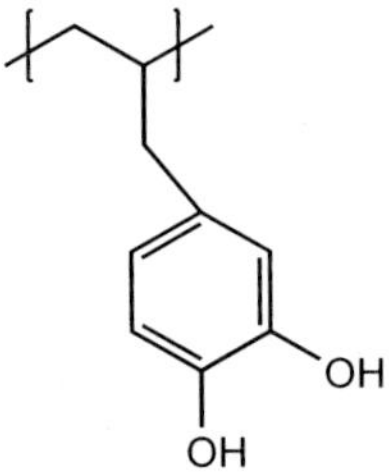

Scheme 12

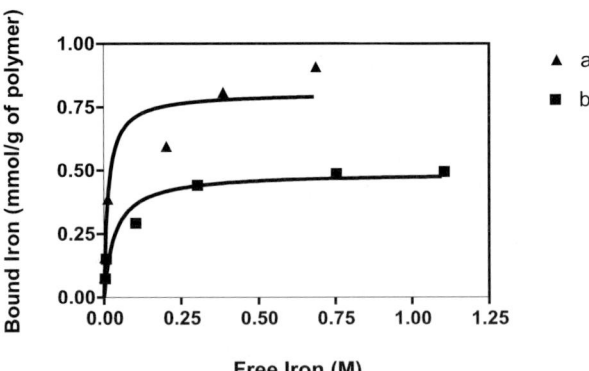

Fig. 5 Iron-binding isotherms of polymeric hydroxamic acid resins at pH: (a) 7.0; (b) 3.5

using rodents have shown that the use of a hydroxamic acid-based hydrogel has arrested intestinal absorption of dietary iron [58]. The polymers were well tolerated by the test animals, indicating their overall biocompatibility.

2.2
Bile Acid Sequestrants: Polymeric Cholesterol-Lowering Drugs

Increased plasma total cholesterol and low-density lipoprotein cholesterol (LDLc) are established risk factors for atherosclerosis, which is the underlying cause of coronary heart disease and most strokes [59]. The reduction of elevated LDLc is one of the most common therapeutic approaches to treat this disease. The majority of individuals at risk of cardiovascular disease require only a modest (20 to 30%) reduction in LDLc level to minimize this serious and often life threatening health risk [60]. HMG-CoA reductase inhibitors (more commonly known as statins) are the most widely used drugs for reducing plasma LDLc and have been shown to significantly reduce the risk of coronary events and strokes. These findings have led to recent guidelines for expanding the use of cholesterol lowering drug therapies to more patients at risk of cardiovascular disease [61]. An estimated 36 million people in the United States alone are candidates for cholesterol lowering drug therapy.

Despite the clinical success of statins, there is still a need for alternative therapies to reduce blood LDLc. For example, statins are not recommended for pregnant women. They are also not recommended for pediatric use or for patients suffering from liver disease. Furthermore, some patients do not achieve the LDLc goal with statin therapy alone. There are also long-term potential safety issues associated with statins such as liver dysfunction and musculoskeletal symptoms. Since cholesterol lowering therapies are generally life-long, these safety factors are significant for such an extended treatment. This has been evident from the recent withdrawal of a statin, cerevastatin (Baycol), from the market as a result of several cases of rhabdomyolysis leading to death [62, 63].

The molecular regulation of cellular cholesterol metabolism has been elucidated by Brown and Goldstein [64]. Cholesterol is synthesized in the liver by the enzyme HMG-CoA reductase. Subsequently, it is transformed into bile acid in the liver and secreted to the gall bladder. The statins inhibit HMG-CoA reductase, which is a key rate-limiting enzyme in cholesterol biosynthesis. The effective removal of the bile acid from the bile pool is another viable approach to reduce plasma LDLc. Removal of bile acid from the body results in upregulation of bile acid biosynthesis, which subsequently leads to a corresponding overall drop in plasma cholesterol levels [65]. The biochemical process of cholesterol metabolism is schematically illustrated in Fig. 6.

Bile acid sequestrants (BAS) are crosslinked polymeric cationic gels that bind anionic bile acids in the GI tract and subsequently eliminate them from

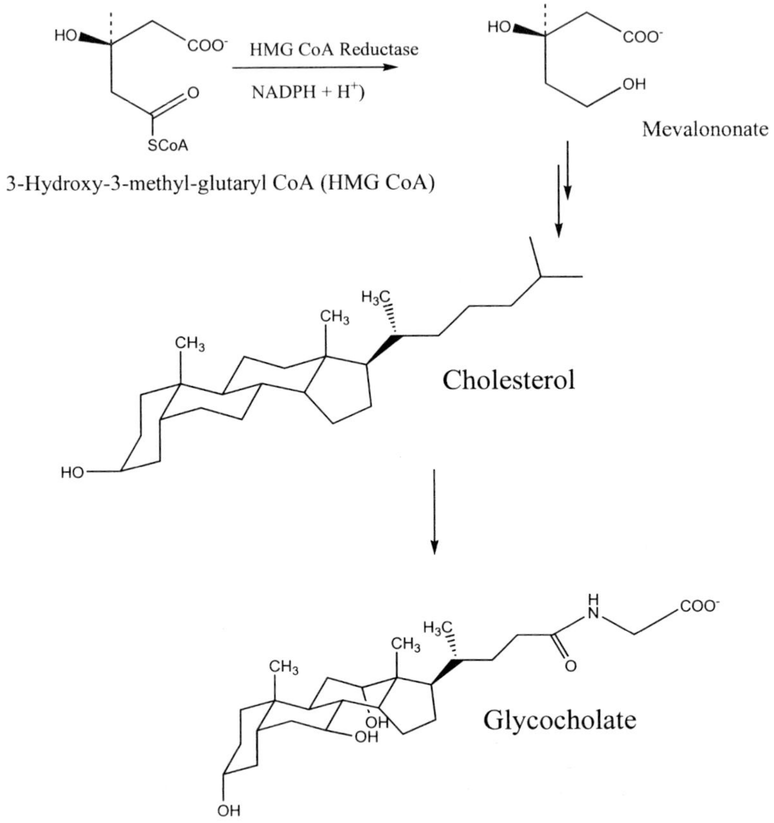

Fig. 6 Biosynthesis of cholesterol and its metabolism into bile acid

the body along with the feces [66]. The use of these polymeric gels for sequestering bile acids is indeed an established approach for treating patients with elevated plasma cholesterol [67]. Being non-absorbed, these polymeric drugs do not exhibit the systemic side effects that are associated with statins, and the BAS have over 30 years of clinical experience with a good safety record. Until recently, two cationic polymers, namely cholestyramine (Scheme 13) and colestipol (Scheme 14) have been the only approved bile acid sequestrants on the market. Despite the appeal of their safety profiles, these two first generation bile acid sequestrants have seen decreased use since their introduction to the market. While these two BAS exhibit high in vitro capacity, they have shown low clinical potency. For example, the doses required for a 20% cholesterol reduction with cholestyramine and colestipol are typically 16 to 24 g/day [68]. The requirement of this high daily dose led to reduced patient compliance and hence limited use of these first generation BAS in clinical settings.

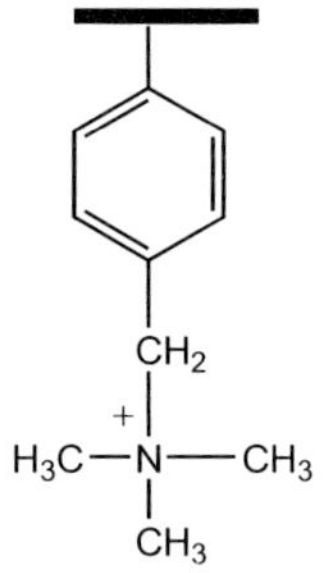

Scheme 13

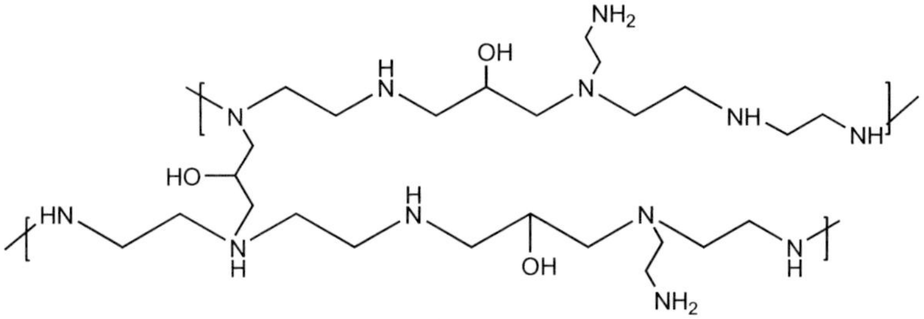

Scheme 14

The low in vivo efficacy of BAS has been ascribed to competing forces of the active bile acid transporter system of the GI tract, which the BAS has to encounter to strongly hold on to the bile acid from systemic reabsorption [69]. It appears that, for a polymer to be a potent BAS, it must have high binding capacity, strong binding strength, and selectivity towards bile acids (over other anionic and amphiphilic species in the GI tract) in the presence of competing desorbing forces of the GI tract. Therefore, a potent BAS needs to exhibit slow off-rates of bound bile acids from the polymer resin to effectively overcome the active transport of bile acids from the GI tract. The design rules for making potent BAS need to take into consideration the physicochemical features of bile acids associated with their structures (Scheme 15). A typical bile acid possesses an anionic group and a hydrophobic core, which are responsible for their biological detergent properties. Therefore, while electrostatic interaction is the primary force required for bile acids to complex with cationic polymers, a second attractive force that needs to be considered is the hydrophobic interaction between the sequestrant and the bile acid. Moreover, favorable swelling characteristics of these cationic hydrogels in physiological environments are required for attaining high capacity (that would make use of maximum binding sites in the polymers). Thus, a balanced combination of hydrophilicity (high capacity) and hydrophobicity (to slow down the rate of desorption), along with an optimum density of cationic groups would constitute key features of potent BAS [70].

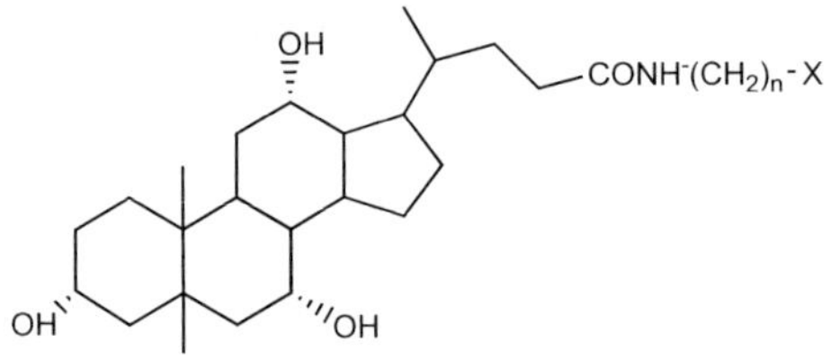

15a: n =1 X = COOH; **15b:** n =2, X = SO$_3$H

Scheme 15

Careful consideration of these desired features has led to the discovery of a number of new generation bile acid sequestrants over the last decade from our laboratories as well as from other groups. Thus, a large body of literature (that has been published over the years) is now available [71–75]. Structural features of some of the representative new bile acid sequestrants are summarized in Table 3. In general, the common features of these cationic hydrogels are the presence of amine/ammonium groups containing a whole range of substituents around the nitrogen atom. Additional structural features of these polymers include the presence of hydrophobic chains. Different kinds of polymer backbones including vinyl and allyl amine polymers, (meth)acrylates, (meth)acrylamide, styrene, carbohydrates, polyethers, and other condensation polymers have been considered [76–78]. The effect of polymer chain architectures (such as block copolymer) on bile acid sequestration has also been evaluated [79]. While a large number of polymers have been synthesized

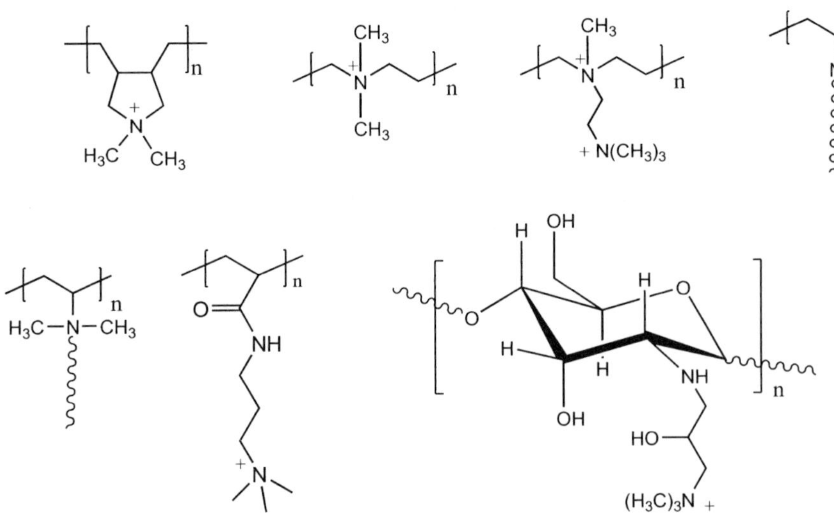

Table 3 Chemical structures of some representative bile acid sequestrants

and tested in preliminary in vitro and in vivo studies, very few of these polymers have entered preclinical development and subsequent human clinical trials. Some promising BAS that have entered the clinical trial include: DMP-504 (Scheme 16), colestimide (Scheme 17), SK & F 97426-A (Scheme 18), and colesevelam hydrochloride (Scheme 19). The key structural features of these polymers are an optimum combination of charge density, hydrophobic tails,

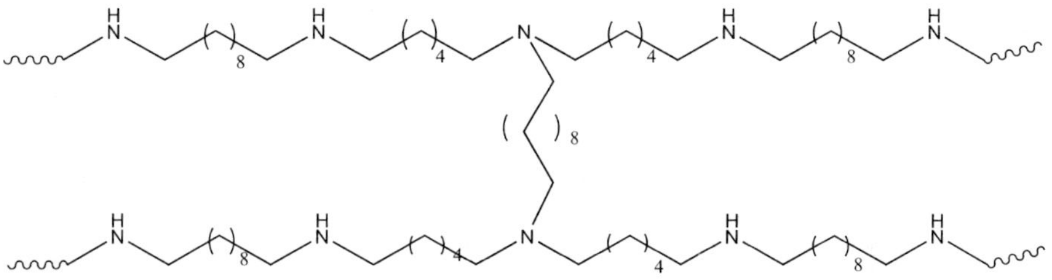

Scheme 16

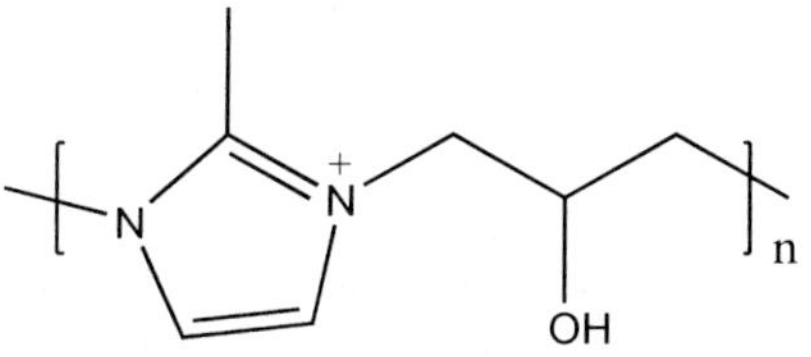

Scheme 17

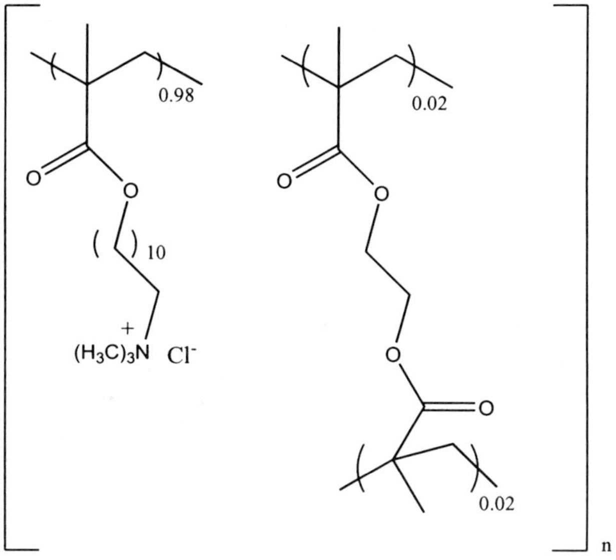

Scheme 18

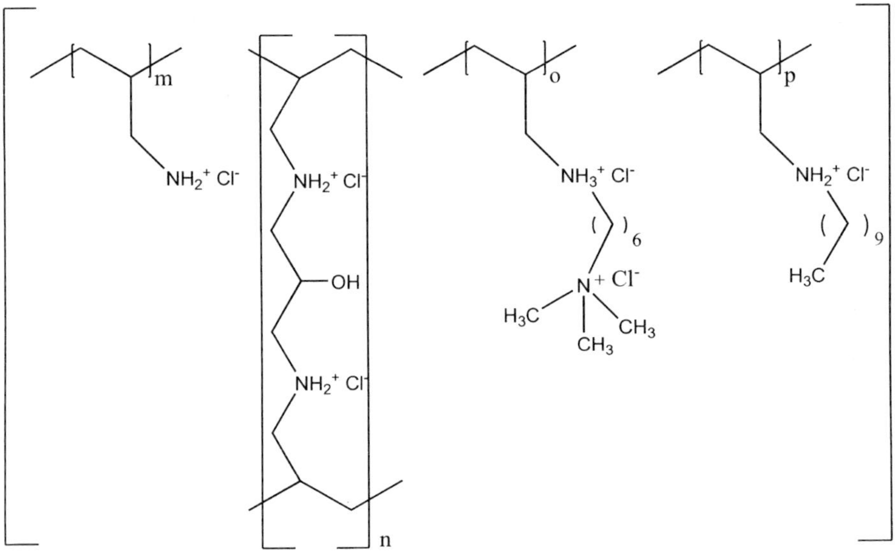

Scheme 19

and water swelling properties [80, 81]. From this new generation of bile acid sequestrants only two compounds have been approved by the regulatory agency for marketing. Colestimide has been approved for marketing in Japan and is sold under the trade name Cholebine [82]. Colesevelam hydrochloride has been approved for marketing in the United States and is being sold under the trade name WelChol [83]. Both polymers exhibit lower rates of side effects and are better tolerated than previously marketed BAS. These BAS have been recommended for use as monotherapy or in combination therapy by co-administration with other cholesterol lowering drugs, such as statins.

3
Polyvalent Interactions and Anti-Infective Polymeric Drugs

Polyvalent interactions are characterized by the simultaneous binding interaction between multiple ligands on one molecular entity and multiple receptors on another (cells, viruses, proteins, etc). These phenomena are frequently encountered in biological systems. Due to a multitude of chelating effects, polyvalent interactions can be significantly stronger than the corresponding monovalent interactions. Polyvalent interactions form the initiation steps for a large variety of key biological processes such as cell-surface or receptor-ligand recognition events. They can provide the basis for mechanisms of both agonizing and antagonizing biological interactions that are fundamentally different from those available in monovalent systems. A schematic illustration of the principle of polyvalency is presented in Fig. 7. An elaborate description of the underlying theory of polyvalency and its importance in biological processes has been described by Whitesides and coworkers in a recent review article [84].

Recognition of this important interaction in biological systems is the basis of a new paradigm for the design and discovery of human therapeutics [85]. This approach is particularly relevant when the interaction between a monovalent ligand and a polyvalent receptor is weak, since polyvalency can significantly enhance the binding strength leading to highly potent drug candidates. Thus, a polyvalent agent bearing two or more chemically bound ligands can interact at two or more receptor sites on a target pathogen leading to increased binding strengths resulting in effective inhibition and/or sequestration of the target pathogen of interest. Polymeric systems, wherein a collection of similar or different ligands can be covalently linked together on a single polymer chain provide a unique scaffold to elaborate this concept. In principle, the inherently polyvalent nature of polymeric materials can translate into longer-lasting and more potent therapeutics. Utilization of this new concept of polyvalent ligand-substrate interaction has led in recent years to the discovery of a number of therapeutically relevant polymeric species that have been found to sequester/inhibit pathogenic toxins, viruses, and bacteria. Some of these polymeric drugs have advanced into human clinical trials, further supporting the validity of this novel concept in drug discovery (vide infra).

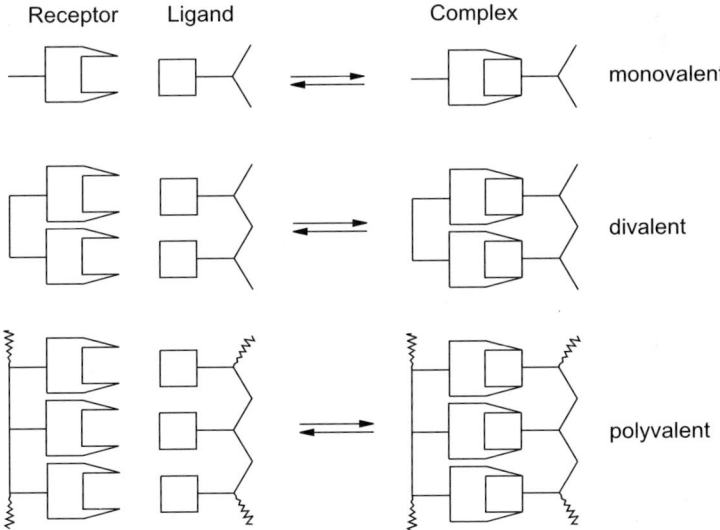

Fig. 7 Schematic illustration of monovalent, divalent, and polyvalent receptors, ligands, and formation of their complexes

3.1
Polymeric Sequestrants of Toxins

Toxins are produced by pathogenic microorganisms in the host body. They are classified into two categories: exotoxins and endotoxins [86, 87]. Exotoxins are generally proteinous materials released by these microorganisms. Endotoxins are lipopolysaccharides and consist of polysaccharide segments and glycolipid segments that constitute the outer cell membranes of all Gram-negative bacteria.

Upon secretion from microorganisms, these toxins can travel within host organism and can cause damage in organs far from the initial site of infection. The pathogenic effects of bacterial toxins can include diarrhea, hemolysis, destruction of leucocytes, paralysis, diarrhea, and septic shock. The outcomes of these events can be life threatening [88]. Some of the microorganisms that release life-threatening toxins in the GI tract include food poisoning organisms like *Staphylococcus aureus*, *Clostridium perfringens*, and *Bacillus cereus*, and the intestinal pathogens like *Vibrio cholerae*, *Escherichia coli*, and *Salmonella enteritidis*. Furthermore, anthrax toxin is produced by *Bacillus anthracis*. In most cases, the causative agents responsible for major symptoms of diseases associated with these pathogens are their associated exotoxins [89, 90].

3.1.1
Sequestration of *Clostridium Difficile* Toxin

Clostridium difficile (*C. difficile*) is responsible for large numbers of episodes of diarrhea that arise from anti-bacterial treatment in nosocomial settings [91]. Since normal colonic flora inhibit the growth of *C. difficile*, the outbreak of this disease has been attributed to the disruption of normal colonic flora by antibiotic treatment, as is prevalent in hospital settings. *C. difficile* releases two high molecular weight proteins (toxins A and B), which are the primary causes of diarrhea in patients with *C. difficile* infection [92, 93]. The traditional approach to treat *C. difficile*-associated diarrhea has been the use of one of two antibiotics: metronidazole or vancomycin. Although the use of antibiotics is often effective in eliminating *C. difficile* infection initially, each repeated use sterilizes the gut, thereby increasing the chance for further infection by *C. difficile*. In many cases, repeated cycles of antibiotic treatment, followed by re-infection by *C. difficile* leads to long hospital stays and patient morbidity. Furthermore, there has been an increasing body of documentation pointing to growing resistance of *C. difficile* to antibiotics [94].

Since disease symptoms are attributed to the toxins released by *C. difficile*, and re-colonization of *C. difficile* is attributed to the inhibition of normal microbial flora growth by antibiotic treatment, selective sequestration and neutralization of these toxins appears to be an attractive and safe antibiotic-free therapy to treat this disease. The proposed approach would treat the infection without disrupting normal bacterial growth in the GI tract. Being proteinaceous substances, bacterial toxins possess multiple binding sites (derived from amino acid side chains). The binding sites can be used as anchoring sites to effectively interact with multifunctional polymers (carrying complementary binding sites) through polyvalent interactions. Some of the early studies to use polymeric ligands to sequester *C. difficile* toxins utilized anion exchange resins like cholestryramine and colestipol (the bile acid sequestrants described above). However, the effectiveness of these polymeric ion-exchangers as sequestrants to bind and remove *C. difficile* toxins was found to be modest at best [95].

Systematic investigations in our laboratories have led to the discovery of polymeric multivalent ligands that effectively bind and neutralize *C. difficile* toxins. From the various polymers evaluated, a series of high molecular weight, water-soluble anionic polymers (Schemes 20–22) have been found to

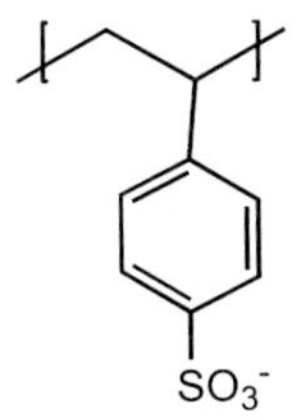

Scheme 20

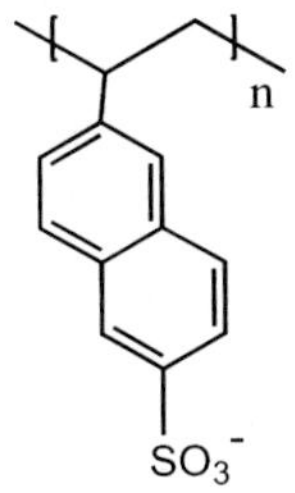

Scheme 21

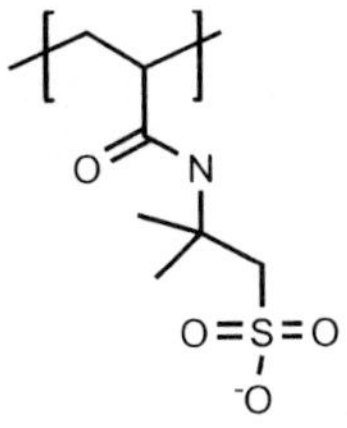

Scheme 22

be the class of polyvalent ligands that are particularly effective in sequestering and neutralizing *C. difficile* toxins [96, 97]. Two important characteristics of these polymers are the presence of sulfonic acid groups and high molecular weight. Careful in vitro studies using pulsed ultrafiltration binding experiments and fluorescence polarization spectroscopy revealed that the binding constants of complexes between one of these polymers and toxins A and B were 133 nM and 8.7 µM, respectively [98]. The ability to bind these toxins is not a general feature of all anionic polymers. For example, while the sodium salt of poly(styrene sulfonic acid) (Scheme 20) effectively binds both toxins, poly(sodium 2-acrylamido-2-methyl-1-propanesulfonate) (Scheme 22), a high molecular weight polyanion of similar charge density does not bind either of the toxins to any measurable extent. This implies that the chemical characteristics of monomer units that make up these polyvalent ligands influence the toxin-binding properties of the polymer. Furthermore, the binding strength was found to be dependent on molecular weight, with lower molecular weight polymers exhibiting very low binding strength. From these observations it is evident that sequestration of *C. difficile* toxins by polyanions is not purely electrostatic in origin. On the other hand, it appears that the sodium salt of poly(styrene sulfonic acid) and related anionic polymers interact with toxins A and B through a multitude of weak interactions. Amplification of these individual interactions, as a result of polyvalency, translates into a substrate-binding event of very high binding strength. From fluorescence polarization data, it was estimated that one molecule of toxin A interacts with ∼ 800 monomer units present in a polymer chain. Thus, a single polymer chain of molecular mass of 300 kDa can wrap around a toxin molecule ∼ 3–4 times [98]. This suggests that for effective toxin binding of therapeutic relevance, large polyvalent interactions between the polymer and protein surface is needed [99]. A schematic presentation of the process of multiple contacts between the polymer and toxin molecule is illustrated in Fig. 8. The

Fig. 8 Schematic representation of the sequestration of C. difficile toxin by the polyvalent ligand

in vitro binding activity of these high molecular weight polyanions correlate well with the in vivo biological activities. Thus, the lead polymer prevented the mortality of 80% of hamsters suffering from severe *C. difficile* colitis [100]. These polymers are non-antimicrobial and do not interfere with the activities of standard antibiotics.

High molecular weight sodium salt of poly(styrene sulfonic acid) derivative was selected as the clinical candidate for treating *C. difficile* infection. This compound, under the generic name of Tolevamer, has been successful in both phase I and phase II human clinical trials.

Besides the above anionic polymer, another class of polymers bearing pendant oligosaccharide groups has also been investigated as possible sequestrants for *C. difficile* toxins that have progressed to human clinical trials [101, 102]. The underlying principle behind this polymeric sugar agent for the treatment of *C. difficile* infection is that toxin A has shown a lectin-like activity, which allows it to bind to an oligosaccharide receptor on epithelial cells. Furthermore, toxin B has been found to bind erythrocytes. These findings suggest that the cell invasion/binding of *C. difficile* may be mediated by interactions with cell surface carbohydrate receptors. Therefore, polymers bearing pendant sugar residues may compete with human cells towards *C. difficile* toxin. After identifying oligosaccharide sequences, appropriate oligosccharide molecules that are specific for both toxins were conjugated to different polymer backbone (Scheme 23). Lengths of tethering arms linking the polymer backbones with oligosaccharide moieties were appropriately optimized to maximize the binding strengths between polymeric ligands and toxins. The oligosaccharide sequences that were found to improve toxin binding include maltose, cellobiose, isomaltotriose, and chitobiose. Polymer carriers examined in this study include substituted polystyrene and other polyolefin backbones. Inorganic carriers such as biogenic silica and kaolinite were also used. These polymeric toxin binders (under the trade name of SYNSORB) were found to be effective in neutralizing *C. difficile* toxins and in controlling

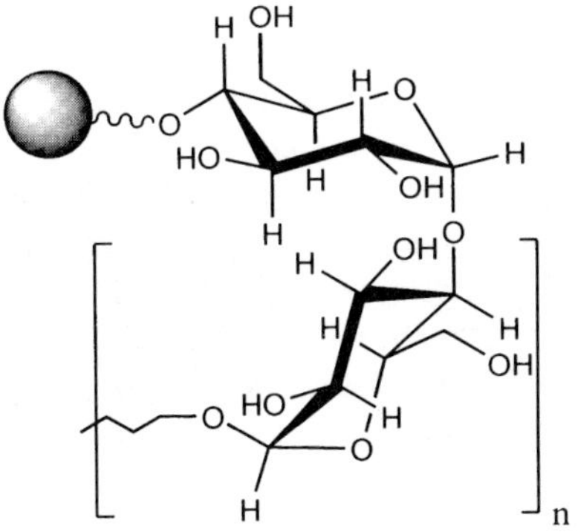

Scheme 23

diarrhea in animal models. One of these polymeric carbohydrate agents had also entered human trials. However, due to lack of therapeutic efficacy, it was withdrawn after a phase II clinical trial.

3.1.2
Sequestration of Anthrax Toxin

Anthrax toxin is produced by the Gram-positive bacteria *Bacillus anthracis* and is the primary cause for the major symptoms of the disease [103]. In general, the occurrence of anthrax in clinical settings is rare. However, growing concern over bioterrorism and biological warfare involving *B. anthracis* in recent years has put the effort to discover and develop anti-anthrax agents on a high-priority list. Although vaccines against anthrax exist, several factors make mass vaccination difficult. While treatment with antibiotics could eradicate bacteria from the host, the continuing action of the toxin in inducing damage after symptoms have become evident makes antibiotic treatment of limited clinical benefit. Therefore, development of agents that can sequester anthrax toxin (thereby inhibiting its action) is an attractive adjunct to antibiotic therapy for treating anthrax infection.

Like most intracellularly acting toxins (such as ricin and botulinum neurotoxin), anthrax toxin consists of two subunits: an activating region (subunit A) and a promoter region (subunit B). The subunit B interacts with the cell surface receptor that is specific for the toxin. However, the bacterium secretes three separate proteins: a single receptor-binding moiety termed protective antigen (PA), and two enzymatic moieties, termed edema factor (EF) and lethal factor (LF). Upon release from bacteria as non-toxic monomers, these three proteins combine and undergo a cascade of processes through the formation of a cell bound heptameric fragment called PA63, which finally attacks the macrophages [104]. The heptameric PA63 moiety binds to host cells via a polyvalent interaction of very high binding constant ($K_d \sim 1$ nM).

The biological pathway leading to anthrax infection (as described above) suggest that development of a polyvalent ligand that would compete with PA63 could be a potential agent to inhibit binding of anthrax toxin to the surface of the host cell. While the overall structural requirements for binding of PA63 heptamer are yet to be investigated, some structural studies have shown the importance of hydrophobic interactions involving Tyr and Trp residues as well as H-bond donor/acceptor sites [105]. By utilizing phage-display library screening, a dodecameric peptide sequence (P1) was identified that binds to PA63 and thus interferes with its interaction with host cells. In a cell culture assay, this peptide exhibited modest potency ($IC_{50} \sim 150\,\mu$M). By attaching multiple copies of this peptide P1 to a polyacrylamide-based polymer backbone, the corresponding polyvalent ligand was obtained. This polymeric ligand contained, on average 22 P1 units and ~ 900 acrylamide units. The IC_{50} value of this polymer in inhibiting the binding of PA63 to host cells was found to be 20 nM. This value corresponds to a nearly 7500-fold increase in potency on a per-peptide basis relative to the free P1 [106]. Optimization of this class of polyvalent ligands by increasing the P1 concentration in the polymer backbone and incorporation of additional hydrophobic groups and H-bond donor/acceptor sites are reportedly under investigation to further enhance the potency of this class of polymers to treat anthrax infection [107]. One of these polymeric agents was evaluated in vivo in Fisher 344 rat models that were intoxicated with anthrax lethal toxin. The polymer, when dosed intravenously, delayed the symptoms and eliminated the toxicity at a dose of 12 nmol equivalent of peptide. No obvious toxicity associated with i.v. administration of the polymer was observed in the test animals during the week-long treatment [106]. The efficacy of this polyvalent ligand in blocking the action of anthrax toxin in vivo suggests that this approach could be useful in developing therapeutically relevant agents to combat possible future risks of bioterrorism involving anthrax toxin.

3.2
Polyvalent Ligands as Antiviral Agents

Viral infection is initiated by the attachment of viruses to specific cellular receptors. This virus-cell receptor attachment process is mediated by viral attachment proteins (VAPs) [108]. It has been postulated that the presence of synthetic analogs of cellular receptors would compete with VAPs towards forming the receptor-binding epitopes of VAPs. It has been calculated that in general, attachment of viruses to cell surfaces involves the interaction of a virus species with at least 10 monomeric cellular receptor units organized into polymeric cellular receptor units [109]. The characteristic feature of such polyvalent cooperative binding is the amplification of multiple low affinity individual contacts to a binding event of very high association constant. This suggests that by presenting polymeric ligands bearing multiple copies of the cellular receptor analogs, competitive inhibitors for viruses can be developed. This design principle has been utilized to develop prototype antiviral agents against viruses such as influenza and rotavirus.

3.2.1
Inhibition of Influenza Virus

Influenza virus A is the primary causative agent responsible for serious cases of human influenza. The influenza infection is initiated by attachment of the virus to the mammalian cell membrane through a process known as hemagglutination. The hemagglutination process is a multivalent interaction between trimers of hemagglutinin (a carbohydrate binding protein present on the viral surface) with multiple sialic acid groups present on the surface of the mammalian epithelial cell. These sialic acid residues are parts of cell-surface glycoproteins [110]. This biological process of viral invasion of mammalian host cells is schematically shown in Fig. 9. One of the potential strategies to treat influenza virus infection is to block the binding of the virus to mammalian cells by presenting polymers bearing several sialic acid groups as competitors for cell surface ligands [111]. Although individual viral surface

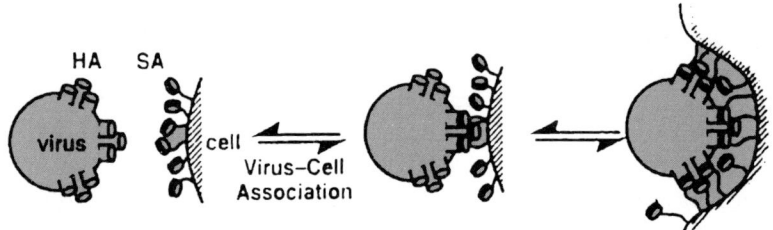

Fig. 9 Schematic illustration of the biological process involving viral invasion of mammalian host cells

Table 4 Representative examples of polyvalent ligands as influenza virus inhibitors

hemagglutinin (HA) binds individual cell-surface sialic acids (SA) weakly, the virus attaches to the cellular surface through multiple interactions between clusters of HA and SA residues. While an individual HA-SA interaction has a K_d of $\sim 2.5 \times 10^{-3}$ M, the interaction between the virus and an erythrocyte has a K_d of $\sim 10^{-12}$ M. In other words, the polyvalent interaction of the latter system is about a billion-fold stronger than the corresponding individual monovalent interaction. In principle, appropriately designed polymeric SA derivatives containing multiple copies of SA may be able to compete with and disrupt the strong virus-cell binding. The research groups of Whitesides et al., Bovin et al., and Roy et al. investigated this phenomenon quite systematically by designing and synthesizing an array of polymers containing side chain SA groups [112–115]. Some of these polymers have shown very impressive affinity enhancement towards influenza virus. The chemical structures of some of the representative polymers are summarized in Table 4. The most effective inhibitor among these polymers is a linear polyacrylamide derivative containing, on a side chain, the C-glycoside of SA (Scheme 24). This polymeric ligand prevents hemagglutination at a concentration of 35 pM, while the corresponding monomeric α-methylsialoside is a very weak inhibitor that inhibits the hemagglutination process at a concentration greater than 2 mM [116].

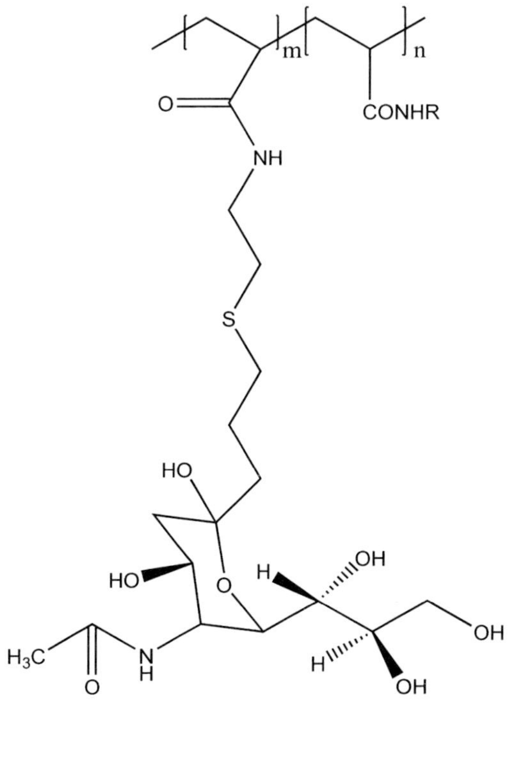

Scheme 24

This is probably the highest increase in potency for any polyvalent system to inhibit virus-erythrocyte binding. A systematic investigation to elucidate the mechanism of this polyvalent interaction between influenza virus and polymeric ligands has shown that a balanced combination of several factors including high affinity multiple ligand density, steric stabilization, and entropically driven enhanced binding contribute to the overall effectiveness of these polymeric inhibitors to prevent cell surface attachment of viruses [117].

Although a great deal of in vitro studies have been carried out, in vivo assays exhibiting the success of these polyvalent ligands in protecting animals against influenza infection have yet to be demonstrated. The success of this approach to develop antiviral agents for influenza would also depend on their favorable toxicity as well as biological activity profiles.

3.2.2
Anionic Polymers as Anti-HIV Agents

The antiviral properties of anionic polymers have recently received a lot of attention as agents to protect against infection with sexually transmitted diseases. Due to the cationic nature of most viruses, several anionic polymers are known to bind viruses. As early as the 1960s, researchers had studied the anti-viral properties of a variety of synthetic polymers [118]. However, not all anionic polymers inactivate viruses. Several classes of anionic polymers have been studied for their ability to inactivate the HIV virus. These polymers include poly(styrene-4-sulfonate), 2-naphthalenesulfonate-formaldehyde polymer, and acrylic acid-based polymers. Certain chemically modified natural polymers (i.e., semisynthetic) such as dextrin/dextran sulfates, cellulose sulfate, carrageenan sulfate, and cellulose acetate phthalate have also been investigated for this purpose. Of a number of such anionic polymers that have shown in-vitro and in vivo anti-HIV activity, a couple of polymeric drug candidates have proceeded to early stage human clinical trials for the evaluation of safety/tolerability [119]. While most of these have shown the desired tolerability and safety, further clinical trials are necessary to discern the therapeutic benefit and see if anionic polymers will be applicable as anti-HIV therapies.

3.3
Polyvalent Antimicrobial Agents

The emergence of microbial pathogens that are resistant to multiple classes of available antimicrobial agents is becoming a major worldwide public health concern. These multidrug resistant bacteria are ubiquitous in both hospital and community settings [120]. The majority of these strains have been found to carry multiple drug resistance factors. At present, the only effective treatment for multiply resistant bacterial infections is vancomycin. Unfortunately, vancomycin resistance itself is becoming a growing problem. For example, methicillin-resistant *Staphylococcus aureus* (MRSA) strains, vancomycin-resistant *enterococci*, and *amikacin-* and β-lactam-resistant *Kleisiella pneumoniae* are some of the bacterial species that are resistant to vancomycin [121, 122]. This rapid emergence of multidrug resistant bacterial strains and the potential threat they pose to human life means that there is a pressing need to discover and develop novel antibacterial agents to overcome the challenges posed by multidrug resistant pathogens.

The polyvalent ligands as antibacterial agents have been considered to exhibit several potential advantages over monomeric antimicrobial agents. Cluster effects from polyvalent ligands would lead to amplification of weak non-covalent bonding interactions between the bacterial surface receptors and the polymeric ligands. Aggregation and precipitation of bacteria by polyvalent ligands is potentially another favorable feature. Finally, polyvalent ligands utilizing multi-point attachments could enhance lysis of the bacterial cell membrane/wall.

Similar to viral infection, most bacterial infections are initiated by adhesion of microorganisms to the mucosal surfaces of the host, mediated in part by bacterial protein adhesins [123]. These adhesins interact with carbohydrate determinants of host cell glycolipids or glycoproteins. This underlying mechanism for bacterial infection suggests that appropriate polyvalent sugar derivatives could competitively block the attachment of microbial adhesin to the host mucosal surface resulting in protection against infection. This concept has been explored through the synthesis of a number of polymers bearing acid-functionalized glycoside moieties. Olefinic monomers containing glycoside moieties and acid functional groups such as O-sulfo and O-carboxymethyl groups were prepared and converted to various copolymers. These polymers were found to be effective in vitro against a number of bacterial targets [124].

A second approach to design polyvalent ligands as antimicrobial agents based on cationic polymers has also been systematically explored in our laboratories. The mechanism of the antimicrobial action of these polymers has been attributed to their enhanced ability for cell lysis. These polymers were designed as mimetics of certain cationic amphiphilic peptides containing multiple arginine and lysine residues. These cationic peptides, which are known as antimicrobial peptides (or defensins) cause cell lysis. The cell lysis is mediated by the interaction of the positive charges of the peptides with negative phosphate head groups of cell membrane phospholipids [125]. A series of amphiphilic cationic polymers were prepared bearing amine and quaternary ammonium groups as well as hydrophobic tails as defensin analogs. These polymers were found to exhibit antimicrobial activity against a number of microbes [126]. In particular, some of these polymers (Schemes 25–27) were found to be very effective against *Cyptosporadium parvum* (*C. parvum*). Until the advent of therapeutic HIV protease inhibitors, *C. parvum* was a pri-

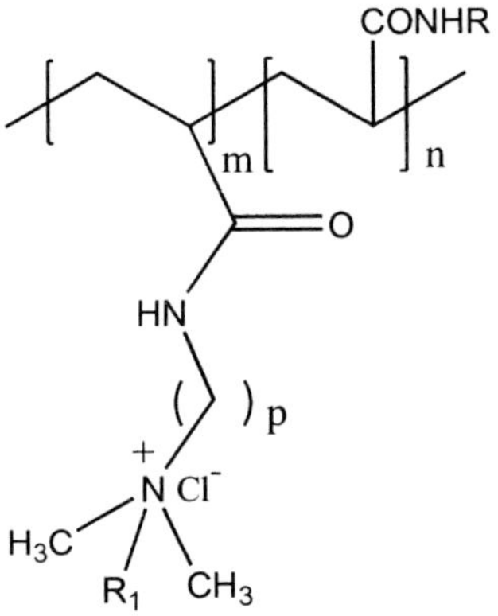

Scheme 25

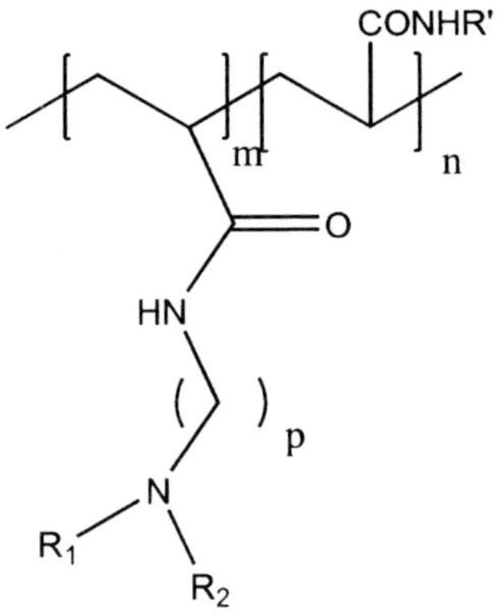

Scheme 26

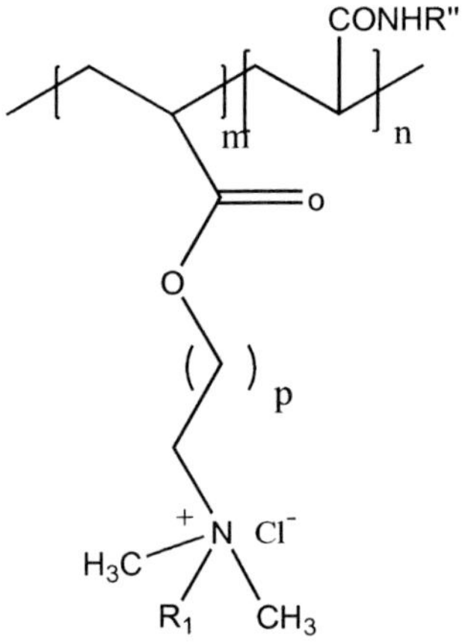

Scheme 27

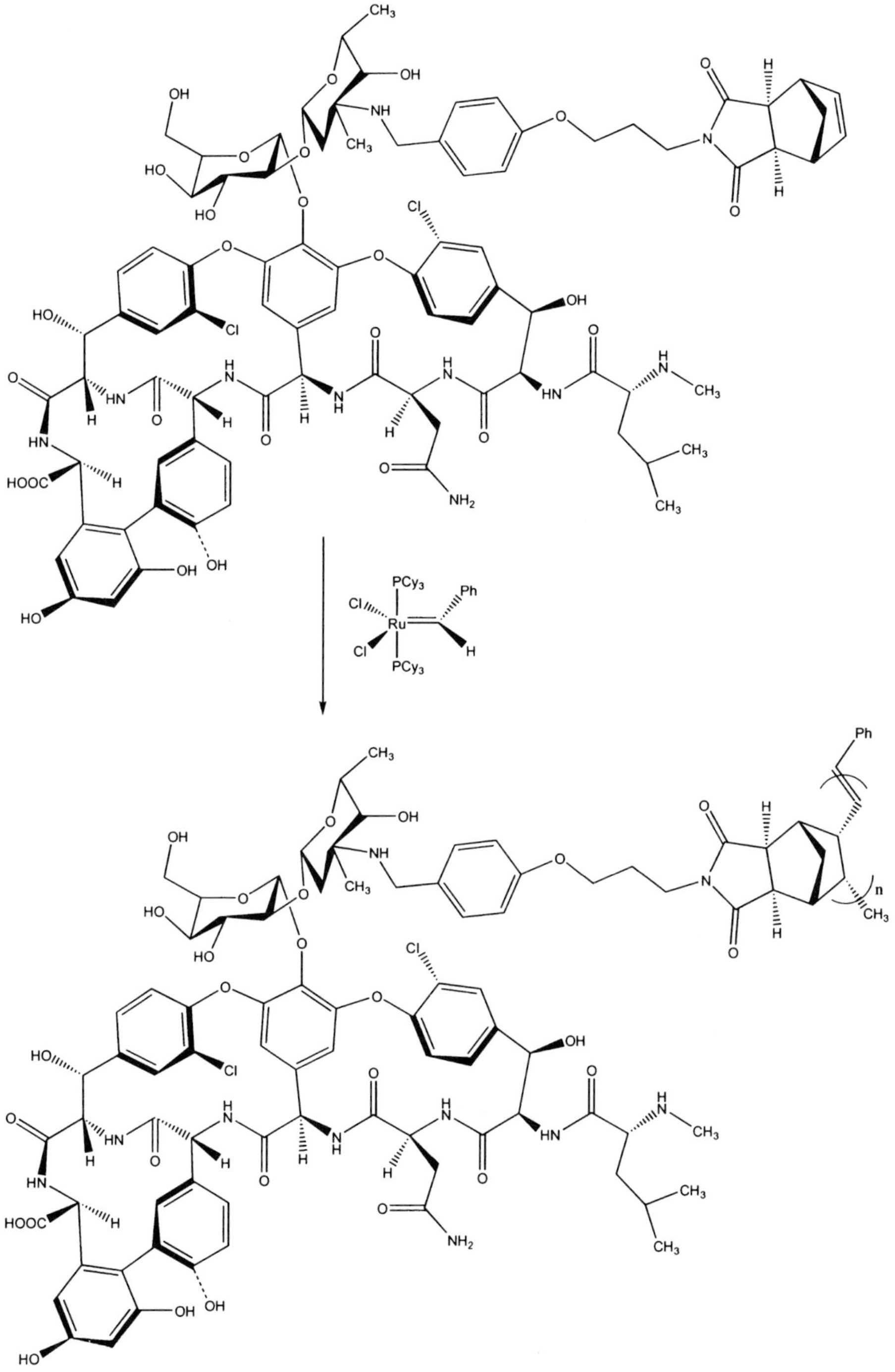

Fig. 10 Synthesis of polyvalent vancomycin

mary target for drug discovery to treat GI tract infections in individuals with HIV infection [127]. In an in vivo study, some lead polymers from this series of polycations were found to be superior to the commonly prescribed antibiotic, paromomycin [126].

Synthesis of a multivalent vancomycin derivative exhibiting significantly enhanced antibacterial activity against vancomycin-resistant *enterococi* (VRE) has been recently reported [128, 129]. This work was built upon the finding that dimeric vancomycin showed enhanced affinity towards the L-lysyl-D-alanyl-D-alanline peptidoglycan precursor that is responsible for the growth of the bacterial cell wall [130]. A polymeric vancomycin derivative was obtained by ring opening metathesis polymerization (ROMP) of a functional cyclic olefin monomer containing the vancomycin moiety (Fig. 10). The antimicrobial property of the monomer (Fig. 10 top) was similar to native vancomycin. Upon incorporation into the polymer backbone, the activity of the corresponding polymeric vancomycin derivative (Fig. 10 bottom) was found to have increased by nearly 60 fold, when tested against *S. aureus* and *enterococci* [128]. Although no in vivo data is available to date, this promising in vitro result supports the important role that polyvalency can play in the discovery of a new generation of antimicrobial agents.

4
Polymeric Drugs for the Treatment of Autoimmune Diseases

In simplest terms, autoimmune diseases arise when the host immune system mistakenly attacks itself. Ordinarily the immune system uses a number of defense mechanisms to prevent the development of these autoimmune responses by directing T cells (defense cells) to distinguish foreign invaders [131]. Bypassing the protection against autoimmunity leads to inflammation in various parts of the human body. Multiple sclerosis and rheumatoid arthritis (RA) are two of the most common autoimmune diseases [132].

4.1
Polymeric Drugs to Treat Multiple Sclerosis

Multiple sclerosis (MS) is one of the most common inflammatory diseases of the central nervous system associated with immune activity directed against central nervous system antigens. As a result, this disease affects the brain and the spinal cord. Destruction of the regulatory mechanism that guards against autoimmunity leads to inflammation of the central nervous system. This disabling disease affects over 2.5 million people worldwide, particularly young adults [133]. The pathophysiology of MS has been attributed to the infiltration of autoreactive T cells, degradation of myelin basic protein (demyelination), production of excessive inflammatory cytokines such as tumor necrosis factor alpha (TNF-α), etc. Although the knowledge around the immunopathogenesis of MS has expanded over the years, therapeutic advances to treat this debilitating disease have been modest [134]. The mainstays of therapeutic strategies to treat MS include the use of agents that trigger immunosuppression and immunomodulation such as mitoxantrone, recombinant interferon-β (e.g., Avonex®, Humira®), and glatiramer acetate, etc. [135–137].

Glatiramer acetate (GA) is an amino acid-derived synthetic copolymer developed by Sela and cowokers at the Weizman Institute of Science. It acts as an alternative to interferon-β for the treatment of certain forms of MS. The discovery of this polymer represented a significant breakthrough in polymeric drug discovery [138, 139]. GA (Scheme 28) is a random copolymer composed of four L-amino acids (alanine, lysine, glutamic acid, and tyrosine in a molar ratio of 4.2 : 3.4 : 1.4 : 1.0). This copolymer is prepared by the ring opening random copolymerization of the corresponding N-carboxy-α-amino acid anhydrides. Systematic in vitro (using murine T-cell lines) and in vivo (using experimental allergic encephalomyelitis animal models) studies have indicated the beneficial effects of this polymer in treating MS. The possible mode of therapeutic action of GA has been attributed to its ability to compete with immunodominant myelin basic protein (MBP), which is one of the major autoantigens implicated in the pathogenesis of MS [140]. MBP is known to sensitize autoreactive T-cells. The autoreactive MBP-specific T-cells migrate into the central nervous system (CNS) and mediate the pathogenesis of MS. The composition of glatiramer acetate is postulated to resemble a portion of MBP (MBP 85–89). This epitope presented by GA competes with MBP for binding with T-cell receptors (through cross-reaction) leading to antigen specific intervention of the autoimmune process. In human clinical trials glatiramer acetate demonstrated a significant decrease in the number of relapses and rate of progression of the disease. Glatiramer acetate has been approved by FDA and European agencies for the treat-

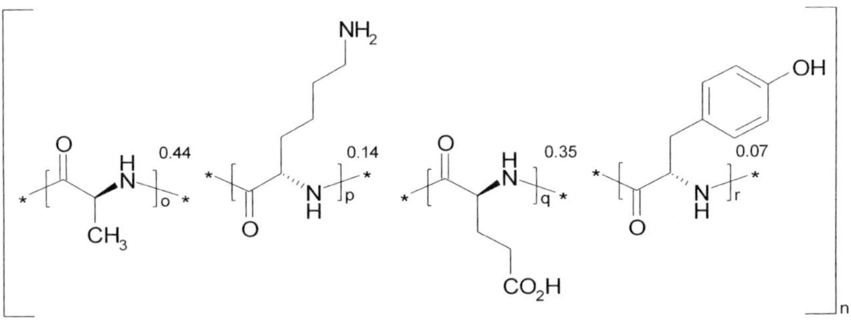

Scheme 28

ment of the relapsing-remitting form of MS in the USA and Europe, and is marketed under the brand name Copaxone by Teva Pharmaceutical Industries [141]. Copaxone is the first synthetic polymeric drug approved for systemic treatment.

Although widely used for the treatment of MS, Copaxone does not completely eliminate the frequency of relapse of the disease. Since the mode of action of GA has been considered to be due to its higher binding strength compared to MBP, the design of a new generation of copolymers based on the anchor residues in MBP as well as by tailoring their mode of arrangement (copolymer sequence) along the polymer chain may lead to more effective immunomodulators. Towards this end, Strominger and coworkers have recently carried out a systematic study to discover a new generation of amino acid-derived immuno-modulating copolymers to effectively treat MS [142–144]. In this investigation, a systematic replacement of tyrosine and glutamic acid with other amino acids were carried out resulting in a series of copolymers with improved binding affinity towards the binding pockets of T-cell receptors. Amongst the various copolymers tested, two copolymers, namely poly(valine-tryptophan-alanine-lysine) and poly(phenylalanine-tyrosine-alanine-lysine) (Schemes 29 and 30) were found to exhibit better activities than GA in a number of in vitro and in vivo

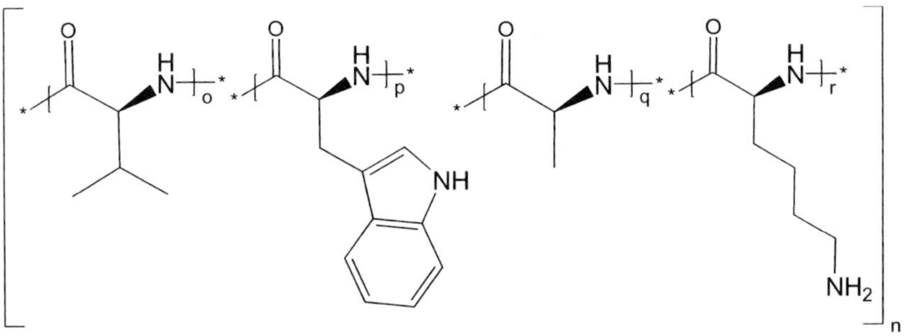

Scheme 29

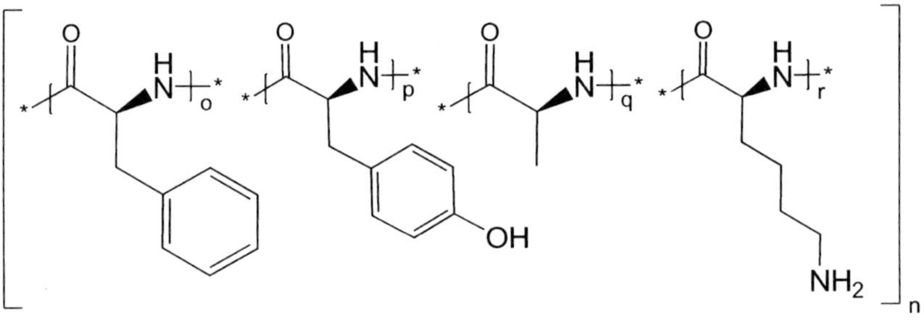

Scheme 30

assays. This new generation of copolymers reportedly produced protective anti-inflammatory cytokines that led to the suppression of the histopathological evidences in EAE animal models. Further development of these promising copolymers for human clinical trials has been undertaken by Peptimmune Corporation [145].

Since GA has been thought to work by antigen-specific intervention against MS, utility of this class of polymeric drugs can be expanded to treat other autoimmune diseases. Hareli and coworkers have recently reported that GA and other related copolymers (containing three amino acids) compete with type II collagen peptide 261–273, a candidate autoantigen in rheumatoid arthritis (RA) for binding to RA associated T-cells [146]. This suggests the possible utility of this class of immunomodulating polymers as a potential therapy for the treatment of RA and other autoimmune diseases.

5
Polymeric Anti-Obesity Drugs

5.1
Obesity and Medical Need

Human obesity has been declared to be one of the most significant health problems in modern times with over 500 million people being overweight. Obesity is associated with an increased risk of developing several serious diseases including hypertension, coronary heart disease, type II diabetes, stroke, osteoarthritis, and cancer. Obesity increases the likelihood of mortality by 20% and recently surpassed smoking as the number one cause of death [147, 148].

There are few medications available for the treatment of obesity. These few anti-obesity agents exhibit modest to minimal efficacy and have been associated with poor side-effect profiles [149, 150]. Thus, there is an urgent need for the discovery and development of new therapeutic agents for the treatment of this significant disease. In general, the strategies to treat/prevent obesity are based on their mechanism of action in maintaining energy balance. When energy intake exceeds expenditure, a state of positive energy balance exists and vice versa. As a result, the role of anti-obesity drugs is to induce negative energy balance until the desired weight loss has been achieved. Besides surgical procedures, the therapeutic approaches to treat obesity fall into four classes: (i) appetite suppressants that act on the CNS by stimulating anorexigenic signals or by blocking orexigenic signals (e.g., sibutramine); (ii) inhibitors of fat absorption that act by inhibiting metabolizing enzymes responsible for the digestion of nutrients (e.g., orlistat); (iii) enhancers of energy expenditure that act by increasing thermogenesis; (iv) stimulators of fat mobilization that act by decreasing *de novo* synthesis of triglyceride [151, 152].

5.2
Inhibition of Lipase to Control Digestion and Absorption of Dietary Fat

Consumption of dietary fat is an important contributor to human obesity. Dietary fats are present mainly as mixed triacyl glycerides (TAG), which comprise one molecule of glycerol and three molecules of fatty acid. The first step in the transportation of fat to the circulation is hydrolysis of TAG to free fatty acids catalyzed by gastric and pancreatic lipases [153]. The hydrolysis process is completed in the small intestine and the hydrolysis products—the free fatty acid and sn-2-monoacylglycerols—are absorbed along the brush border membrane of the small intestine (see Fig. 11). Gastric and pancreatic lipases are the principal lipolytic enzymes in the GI tract that are responsible for the hydrolysis of TAG. Thus, inhibition of lipase enzymes in the GI tract reduces the amount of fat that can be absorbed [154]. Targeting this mechanism to develop anti-obesity drugs appears to offer an inherently safe approach as it involves peripheral targets. Orlistat (also known as tetrahydrolipstatin, Scheme 31) is a potent, specific and irreversible inhibitor of pancreatic and gastric lipases and works by inhibiting the action of lipase enzymes in the stomach and small intestine. The mode of action of orlistat involves the formation of a covalent linkage between the serine hydroxyl group in the catalytic triad and the β-lactone ring of orlistat [155]. In human clinical trials orlistat reduced fat absorption by approximately 30%. The undigested fat is almost exclusively excreted with the feces. At the recommended dose it produces a weight loss of $\sim 10\%$ after one year of treatment. Orlistat has been approved by regulatory agencies for the treatment of obesity and has been marketed by Roche under the trade name

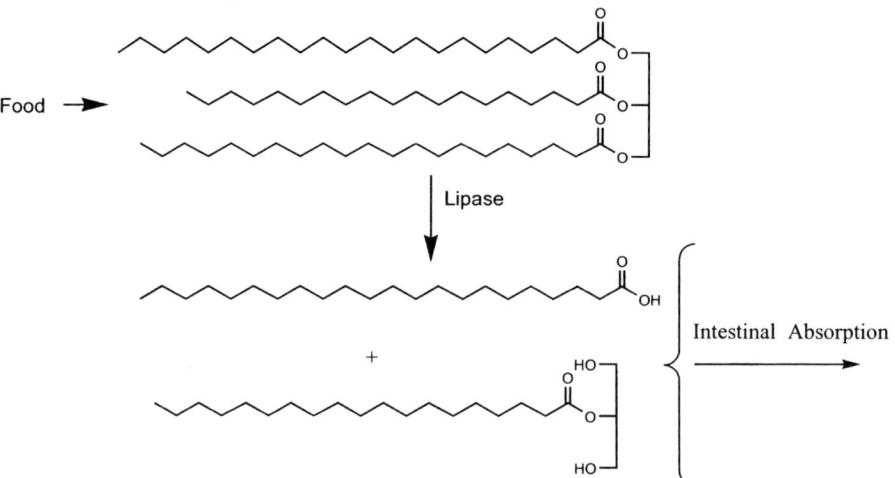

Fig. 11 Lipase-catalyzed hydrolysis of triglyceride and its subsequent intestinal uptake

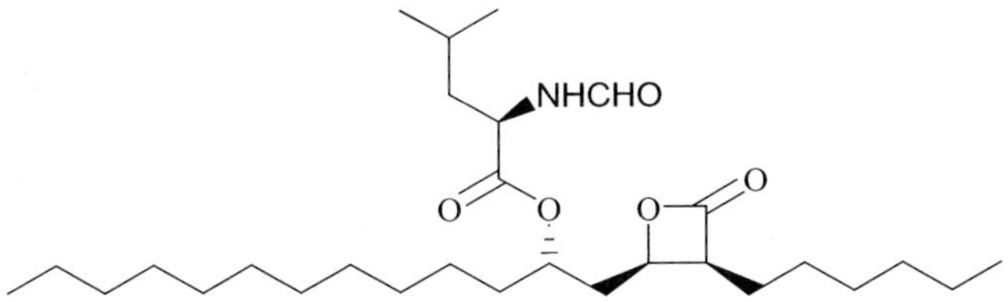

Scheme 31

of Xenical [156, 157]. However, since orlistat blocks fat hydrolysis and increases fecal fat loss, the drug is associated with gastrointestinal side effects that are the direct outcome of TAG malabsorption. These side effects include abdominal pain, steatorrhea, increased flatus, diarrhea, etc. As a result of these side effects, there is decreased compliance with orlistat over time [158].

5.3
Polymeric Fat Binder and Dual Acting Polymeric Lipase Inhibitor-Fat Binder

Besides orlistat, a number of other natural and synthetic inhibitors of human pancreatic and gastric lipases have been identified. Some of these have progressed to clinical development [159]. However, no lipase inhibitor can be expected to overcome the above mentioned side effects. This is due to the fact that as these GI side effects are mechanism related (caused by the presence of non-hydrolyzed fat), options are limited for further improvement in patient compliance by targeting lipases alone. However, a therapeutic intervention that could simultaneously inhibit fat hydrolysis and condense unhydrolyzed fat droplets into a less fluid form may provide a novel approach for lipase inhibitor therapy without the aforementioned side effects. Synthetic functional polymers bearing lipase inhibiting groups and also possessing lipid condensing properties offer a therapeutic opportunity for this purpose. As the first step towards demonstrating the role of synthetic polymers in eliminating the side effects associated with leakage of unabsorbed fat from patients undergoing lipase inhibitor therapy, we adopted a strategy to discover fat-binding polymers that can be co-administered along with a lipase inhibitor, such as orlistat. Dietary fats are usually present in an emulsified form and even the bulk fat in human diets is quickly emulsified in the GI tract. Typically gastric and pancreatic lipases hydrolyze dietary TG at the oil-water interface of a fine emulsion stabilized by physiological emulsifiers such as bile acids and phospholipids secreted by the gall bladder. After initial hydrolysis of some TG in the stomach, the remaining TG and fatty acid move into the small intestine. Here the free fatty acid facilitates the formation of a fine emulsion with particle sizes in the range of one micron. This emulsified lipid phase is efficiently hydrolyzed by pancreatic lipase. The resulting fatty acid is subsequently absorbed by the enterocytes comprising the intestinal wall [160]. However, in the presence of a lipase inhibitor, unhydrolyzed fat passes unchanged through the GI tract. Eventually the emulsifiers and water are absorbed and the triglyceride collects in the lower intestine leading to the oily stool and associated side effects.

Since the unhydrolzyed fat is emulsified in the presence of anionic emulsifiers such as phospholipids, fatty acids, and bile acids, we considered the possibility that non-absorbed cationic polymers could form polyelectrolyte-surfactant complexes with these physiological emulsifiers. The resulting macromolecular network may possess physical properties that could substantially influence the state of the fat emulsion. For example, the polyelectrolyte-surfactant network could provide a rigid or waxy matrix able to encapsulate or stabilize the oil droplets, preventing them from coalescing into a bulk oil phase. In this scenario, the polymeric fat binder could effectively reduce or eliminate the presence of fluid TG in the lower intestine, thereby minimizing the side effects observed for patients undergoing lipase inhibitor therapy for obesity [161]. The proposed rationale for fat binding is illustrated schematically in Fig. 12.

In our laboratories, we prepared a systematic array of polymers of varying hydrophobicity and charge. We devised tests to measure the fat binding efficacy of these materials. This included an in vivo animal model using

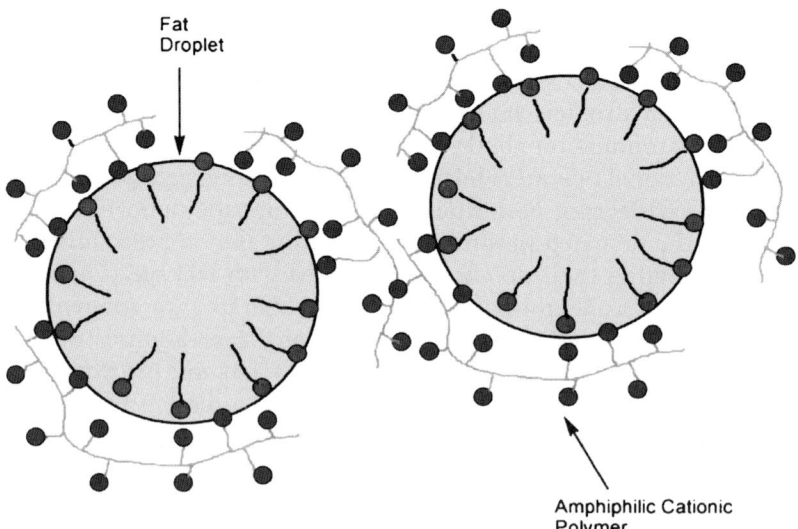

Fig. 12 Schematic representation of sequestration/stabilization of free fat droplets by fat binding polymers

Table 5 Representative examples of fat-binding polymers

rats (fed with lipase inhibitor) to screen polymers for their fat-binding properties. The effect of polymers in overcoming the oily stool side effect (induced by lipase inhibitor treatment) was evaluated with this animal model [162–164].

Although non-ionic polymers such as hydrophobically modified polyethylene glycol and polyethyleneglycol-polypropylene glycol block copolymers are known to exhibit emulsifying properties, they performed very poorly in our animal model. Similarly, copolymers containing anionic and zwitterionic monomers were found to be ineffective. On the other hand synthetic polycations (based on monomers containing amine and ammonium groups) exhibited strong lipid binding properties. A systematic SAR study led to several potent and non-toxic copolymer compositions that completely eliminated the fluid lipid side effect in vivo. Table 5 summarizes a representative list of functional polymer segments that showed favorable fat binding properties and led to overcoming the lipase inhibitor-induced GI side effect [162–164].

In subsequent studies, a novel lipid-binding copolymer was discovered in our laboratory that covalently incorporates a potent and novel lipase inhibitor. This novel copolymer exhibited excellent lipase inhibition in vivo with no fluid-lipid side effects [165]. This polymer has shown safety and efficacy in a number of preclinical studies and has been licensed to Peptimmune, Inc., where it is currently undergoing human clinical trials.

6
Polymer Therapy for Sickle Cell Disease

6.1
Sickle Cell Disease

Sickle cell disease is caused by a mutation in the gene responsible for the production of hemoglobin. Substitution of a single amino acid (valine for glutamic acid) in the sixth position of the B-chain of the hemoglobin molecule results in a hydrophobic region upon deoxygenation [166]. Because the hydrophobic regions aggregate, the abnormal hemoglobin (called hemoglobin S) polymerizes into strands and forms long, rod-like structures. These elongated hemoglobin fibers distort the blood cell, producing the characteristic crescent or "sickle" shape. Surface molecules are also expressed that promote abnormal adhesion of the defective blood cell. Normal red blood cells are smooth, flexible, and shaped like a donut and can pass easily through small blood vessels. The damaged sickle shaped blood cells are stiff and are unable to pass through the blood vessels leading to reduced blood flow and sometimes blockage. The blockage (vaso-occlusion) results in severe, debilitating painful episodes and ultimately to the damage of tissues, and are one of the most common and difficult problems caused by sickle cell disease [167].

6.2
Non-Ionic Surfactant for Treating Sickle Cell Disease

In 1987, a treatment suggestion for vaso-occlusive event was proposed with the demonstration that pluronic F68 was able to improve the filterability and rheology of sickle cells. Pluronic F-68 reduces the endothelial adherence and improves the rheology of liganded sickle erythrocytes [168]. Chemically, pluronics are A-B-A type triblock copolymers containing a segment of polypropylene oxide (PPO) that is sandwiched between two polyethylene oxide (PEO) segments (Scheme 32). Because PPO is a hydrophobic segment, pluronic is able to adsorb to hydrophobic molecules or surfaces while the hydrophilic PEO segments can extend into aqueous phases, these polymers have surfactant properties [169, 170]. The physical properties of pluronics can be modified by changing the PEO and/or the PPO block size. This strategy enables the synthesis of block copolymers that have found a wide range of applications from drug and gene delivery to surface patterning [171].

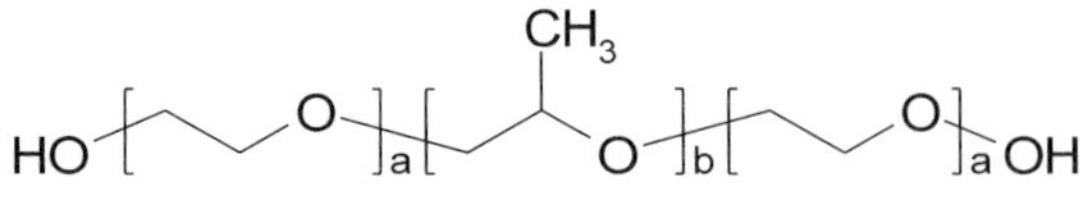

Scheme 32

Stemming from the use of pluronics as solubilizers for small molecule drugs, it became apparent that pluronics is more than just a simple delivery system and provides beneficial biological effects both symbiotically with other treatments and alone.

A purified version of pluronic F68 is being developed by SynthRx under the brand name Flocor as a potential treatment of vaso-occlusive crisis associated with sickle cell disease [172]. After establishing the tolerability of this polymer in clinical studies, further studies of sickle cell patients treated with Flocor showed a reduction in the length of painful episodes and an increase in the number of patients who achieved resolution of the symptoms [173]. The effect observed was significant but small, and the effects seen were more pronounced for children under the age of 15. Further development of Flocor has been reportedly planned, especially for pediatric sickle cell patients.

Although its exact mechanism of action is unknown, Flocor has demonstrated several properties that inherently seem beneficial to sickle cell patients, including: lowering blood viscosity, decreasing red blood cell aggregation, and decreasing friction between red blood cells and vessel walls to increase microvascular blood flow and decrease cell injury. Additional studies would be needed to further clarify Flocor's mechanism(s) of action that may lead to a new generation of polymers for the treatment of sickle cell disease.

7
Conclusions and Outlook

Although polymeric materials have been developed as biomaterials and drug delivery systems, they have not generally been thought of as useful therapeutic agents on their own. In the present article we have attempted to illustrate the extraordinary potential of polymers in the discovery and development of novel human therapeutics. Through appropriate consideration of both disease targets and their mechanisms of action, several functional polymers have been discovered that exhibit useful pharmacological properties. These polymeric drugs capitalize on the unique physicochemical properties of polymer materials and in many cases, exhibit therapeutic properties that cannot be achieved by traditional small molecule drugs. The diversity and activity of these inherently pharmacologically active polymers for the treatment of a number of human diseases that have been either developed or are being evaluated is indeed very impressive. Sequestrants that are confined to the GI tract and carry out their disease-modifying activities are nice examples that support this point. The recognition of polyvalent interactions as a design principle for developing pharmaceutical agents has demonstrated that polymeric drugs can provide a new paradigm for the next generation of pharmaceuticals. Although many have not yet met the goal of exhibiting in vivo biological activity, the successful development of tolevamer suggests that it is not solely an academic curiosity. Finally, our improved understanding of cell biology and biochemistry of various disease targets should further enable us to design polymeric drugs with the desired immunological pharmacological properties for systemic applications. Once these design criteria are identified, exciting and potentially more selective polymer therapeutics will be discovered to treat human diseases, whose medical needs have been either unmet or inadequately met.

References

1. Scheler W (1987) Makromol Chem Macromol Symp 12:1
2. Langer R, Tirrell DA (2004) Nature 428:487
3. Kohn J (1996) Pharm Res 13:815
4. Jagur-Grodzinki J (1999) React Funct Polymers 39:99
5. Ringsdorf H (1975) J Polym Sci Polym Symp 51:135
6. Luo Y, Prestwish GD (2002) Curr Cancer Drug Targets 2:209
7. Ulbrich K, Pechar M, Strohalm J, Subr V (1997) Makromol Chem Macromol Symp 118:577
8. Garnett MC (1999) Crit Rev Ther Drug Carrier Sys 16:147
9. Israel ZH, Domb AJ (1998) Polym Adv Technol 9:799
10. Gros L, Ringsdorf H, Schupp H (1981) Angew Chemie Int Ed Eng 20:305
11. Langer R (1998) Nature 392:5
12. Duncan R (2003) Nature Drug Discovery 2:347
13. Kuo PT, Hopkins FT, Wurzel H (1958) Circulation Research 6:178
14. Regelson W, Holland JF (1962) Clin Pharmcol Ther 3:730
15. Breslow DS (1976) Pure Appl Chem 46:103
16. Chirogos MA, Stylos WA (1980) Cancer Res 40:1967
17. Uglea CV, Panaitescu L (1997) Curr Trends Polym Sci 2:241
18. Bentolila A, Vlodavsky I, Ishai-Michaeli R, Kovalchuk O, Haloun C, Domb AJ (2000) J Med Chem 43:2591
19. Dhal PK, Holmes-Farley SR, Mandeville WH, Neenan TX (2004) Encyl Polym Sci Technol. 3rd Edn. Wiley, New York
20. Johnson LR (1994) Physiology of Gastrointestinal Tract. Raven Press, New York
21. Braun EJ (2003) Comp Biochem Physiol A: Mol & Int Physiol 136A:499
22. Perez GO, Oster JR, Pelleya R, Catalis PV, Kem DC (1984) Nephron 36:270
23. Marrone TJ, Terz KM (1992) J Am Chem Soc 114:7542
24. Gardiner GW (1997) Can J Gastroenterology 11:573
25. Gerstman BB, Kirkman R, Platt R (1992) Am J Kidney Dis 20:159
26. Hsu C (1997) Am J Kidney Dis 29:641
27. Ribeiro S, Ramos A, Brandao A (1998) Nephrol Dial Transplant 13:2037
28. Coburn JW, Norris KC, Sherrard DJ, Bia M, Llach F, Alfrey AC, Slatopolsky E (1988) Am J Kidney Dis 12:171
29. Block G, Hulbert-Shearon T, Levin N (1998) Am J Kidney Dis 31:607
30. Schmidtchen FP, Berger M (1997) Chem Rev 97:1609
31. Lehn JM (1988) Angew Chem Int Ed Engl 27:89
32. Dietrich B, Fyles DL, Fyles TM, Lehn JM (1979) Helv Chim Acta 62:2763
33. Beer PD, Gale PA (2001) Angew Chem Int Ed Engl 40:487
34. Sasaki DY, Kurihara K, Kunitake T (1991) J Am Chem Soc 113:9685

35. Muche M-S, Goebel MW (1996) Angew Chem Int Ed Engl 35:2126
36. Hider RC, Canas-Rodriguez A (1997) US Patent 5698190 (British)
37. Holmes-Farley SR, Mandeville WH, Whitesides GM (1996) US Patent 5496545 (Genzyme Corp)
38. Holmes-Farley SR, Mandeville WH, Whitesides GM (1997) US Patent 5667775 (Genzyme Corp)
39. Holmes-Farley SR, Mandeville WH (2004) US Patent 6726905 (Genzyme Corp)
40. Holmes-Farley SR, MandevilleWH, Ward J, Miller KL (1999) J Macromol Sci Pure Appl Chem A36:1085
41. Chertow GM, Burke SK, Lazarus JM, Stenzel KH, Womboldt D, Goldberg D, Bonventre JV, Slatopolsky E (1997) Am J Kidney Dis 29:66
42. Slatopolsky E, Burke SK, Dillon MA (1999) Kidney Int 55:299
43. Amin N (2002) Nephrol Dial Transplant 17:340
44. Bleyer AJ, Burke SK, Dillon MA (1999) Am J Kidney Dis 33:6694
45. Qunibi WY, Nolan CR (2004) Kidney Int 90:S33
46. Bleyer AJ (2003) Exp Opin Pharmacother 4:941
47. Crichton RR (1991) Inorganic Biochemistry of Iron Metabolism. Ellis Horwood, New York
48. Halliwell B, Gutteridge JMC (1989) Free Radicals in Biology and Medicine. 2nd edn. Clarendon Press, Oxford, UK
49. Boldt DH (1999) Am J Med Sci 318:207
50. Weatherall DJ, Clegg JB (1981) The Thalassaemia Syndromes. 3rd edn. Blackwell Scientific Press, Oxford, UK
51. Cuthbert JA (1997) J Invest Med 45:518
52. Liu ZD, Hider RC (2002) Med Res Rev 26:64
53. Marx JJM (2003) Adv Exper Med Biol 531:57
54. Porter JB, Jawson MC, Huehns ER, East CA, Hazell JWP (1989) Br J Haematol 73:403
55. Badman DG, Bergeron RJ, Brittenham GM (2000) Iron chelators: new development strategies. The Saratoga Group, Florida
56. Miller MJ (1989) Chem Rev 89:1563
57. Mandeville WH, Holmes-Farley SR (1994) US Patent 5487888 (Genzyme Corp)
58. Polomoscanik SC, Cannon CP, Neenan TX, Holmes-Farley SR, Mandeville WH, Dhal PK (2005) Biomacromolecules 6:2946
59. Grundy SM (1998) Endocrinology and Metabolism Clinics of North America 27:655
60. Denke MA, Sempos CT, Grundy SM (1993) Arch Int Med 153:1093
61. (2001) Executive summary of the third report of the "National Cholesterol Education (NCEP) Expert Panel on Detection, Evaluation, and Treatment of High Blood Cholesterol in Adults." J Am Med Assoc 285:2486
62. Rader DJ (2001) Nature Medicine 7:1282
63. Jamal SM, Eisenberg MJ, Christopoulos S (2004) Am Heart J 147:956
64. Brown MS, Goldstein JL (1999) Proc Natl Acad Sci USA 96:11041
65. Grundy SM (1986) In: Fears R (ed) Pharmacological Control of Hyperlipdaemia. Prous Science Publishers, S.A., Barcelona, Spain, p 3
66. Stedronsky ER (1994) Biochim et Biophys Acta 1210:255
67. Mandeville WH, Goldberg DI (1997) Curr Pharm Design 3:15
68. Grundy S (1987) In: Paoletti R (ed) Drugs Affecting Lipid Metabolism. Springer, Berlin Heidelberg New York, p 34
69. Huval CC, Bailey MJ, Braunlin WH, Holmes-Farley SR, Mandeville WH, Petersen JS, Polomoscanik SC, Sacchiro RJ, Chen X, Dhal PK (2001) Macromolecules 34:1548

70. Huval CC, Holmes-Farley SR, Mandeville WH, Sacchiero R, Dhal PK (2004) Eur Polym J 40:693
71. Zarras P, Vogl O (1999) Prog Polym Sci 24:485
72. Mandeville WH, Arbeeny C (1999) Idrugs 2:237
73. Holmes-Farley SR, Mandeville WH, Burke SK, Goldberg DI (2002) US Patent 6423754 (Genzyme Corp)
74. Holmes-Farley SR, Dhal PK, Petersen JS (1998) US Patent 6203785 (Genzyme Corp)
75. Zhang L, Janout V, Renner JL, Uragami M, Regen SL (2000) Bioconjug Chem 11:397
76. Wu G, Brown GR, St-Pierre (1996) Langmuir 12:466
77. Figuly GD, Royce SD, Khasat NP, Schock LE, Wu SD, Davidson F, Campbell GC, Keating MY, Chen HW, Shimshick EJ, Fischer RT, Grimminger LC, Thomas BE, Smith LH, Gillies PJ (1997) Macromolecules 30:6174
78. Huval CC, Bailey MJ, Holmes-Farley SR, Mandeville WH, Sacchiero R, Dhal PK (2001) J Macromol Sci Pure Appl Chem A38:1559
79. Cameron NS, Eisenberg A, Brown GR (2002) Biomacromolecules 3:116
80. Hainer JW, Hunninghake DB, Benedek IH, Broyles FE, Garner DM, Jenkins RM, McGinn A, Pieniazek HJ, London E, Gillies PJ (1997) Drug Dev Res 41:76
81. Wong NN Colesevelam (2001) Heart Disease 3:63
82. Gaudilliere B, Bernardellli P, Berna P (2001) Ann Rep Med Chem 36:293
83. Bays H, Dujovne C (2003) 4:779
84. Mammen M, Choi SK, Whitesides GM (1998) Angew Chem Int Ed 37:2754
85. Borman S (2000) Chem Eng News 78:48
86. Henderson B, Wilson M, Wren B (1997) Trends in Microbiol 5:454
87. RauchhausM, Coats AJS, Anker SD (2000) Lancet 356:930
88. Beutler B, Poltorak A (2001) Critical Care Medicine 29(7 Suppl):S2
89. Pollack M (1983) Rev Infect Dis 5:S979-984
90. Coyle EA (2003) Pharmacotherapy 23:638
91. Bartlett JG (1994) Clin Infec Dis 18:S265
92. KellyCP, LaMont T (1998) Annu Rev Med 19:375
93. Spencer RC (1998) J Antimicrob Chemther 41 (Suppl C):5
94. Alcantara CS, Guerrant RL (2000) Curr Gastroenter Rep 2:310
95. Taylon NS, Bartlett JG (1980) J Infect Dis 141:92
96. Bacon-Kurtz C, Fitzpartick R (2001) US Patent 6290946 (Genzyme Corp)
97. Fitzpatrick R, Huval CC, Bacon-Kurtz CB, Mandeville WH, Neenan TX (2001) US Patent 6290947 (Genzyme Corp)
98. Braunlin W, Xu Q, Hook P, Fitzpatrick R, Klinger JD, Burrier R, Kurtz CB (2004) Biophys J 87:534
99. Kiessling LL, Gestwicki JE, Strong LE (2000) Curr Opin Chem Biol 4:696
100. Kurtz CB, Cannon EP, Brezzani A, Pitruzzello M, Dinardo C, Rinard E, Acheson David WK, Fitzpatrick R, Kelly P, Shackett K, Papoulis AT, Goddard PJ, Barker RH, Palace GP, Klinger JD (2001) J Antimicrob Chemther 45:2340
101. Nilsson UJ, Heerze LD, Liu YC, Armstrong GD, Palcic MM, Hindsgaul O (1997) Bioconj Chem 8:466
102. Heerze LD, Armstrong GD (2002) US Patent 6358930
103. Dixon TC, Meselson M, Guillemin J, Hanna PC (1999) New Engl J Med 341:815
104. Duesbery NS (1998) Science 280:734
105. Benson EL, Huynh PD, Finkelstein A, Collier RJ (1998) Biochemistry 37:3941
106. Mourez M, Kane RS, Mogridge J, Metallo S, Deschatelets P, Sellman BR, Whitesides GM, Collier RJ (2001) Nature Biotechnology 19:958

107. Gujraty K, Kane RS (2004) AIChe Annual Meeting, Austin, Texas, Poster number 568d
108. Bukrinskaya AG (1982) Adv Vir Res 27:141
109. Matrosovich MN (1989) FEBS Letters 252:1
110. Wiley DC, Skehel JJ (1987) Annu Rev Biochem 56:356
111. Weinhold EG, Knowles JR (1992) J Am Chem Soc 114:9270
112. Mammen M, Dahmann G, Whitesides GM (1995) J Med Chem 38:4179
113. Choi SK, Mammen M, Whitesides GM (1997) J Am Chem Soc 119:4103
114. Roy R, Zanini D, Meunier SJ, Romanowska A (1993) Chem Commun 213
115. Matrosovich MN, Mochalova LV, Marinina V, Byramova NE, Bovin NV (1990) FEBS Letters 272:209
116. Mammen M, Dahmann G, Whitesides GM (1995) J Med Chem 38:4179
117. Choi SK, Mammen M, Whitesides GM (1996) Chem Biol 3:97
118. Feltz ET, Regelson W (1962) Nature 196:642
119. D'cruz OJ, Uckun FM (2004) Curr Pharm Design 10:315
120. Kotra LP, Golemi D, Vakulenko S, Mobashery S (2000) Chem Ind 10:341
121. Pearson H (2002) Nature 418:469
122. Walsh C (2000) Nature 406:775
123. Reid G, Sobel JD (1987) Rev Infect Dis 9:470
124. Mandeville WH, Garigapati VR (1997) US Patent 5700458 (Genzyme Corp)
125. Maloy WL, Kari UP (1995) Biopolymers 37:105
126. Mandeville WH, Neenan TX, Holmes-Farley SR (1998) US Patent 6034129 (Genzyme Corp)
127. Matukaitis JM (1997) J Commun Health Nurs 14:135
128. Arimoto H, Nishimura K, Kinuki T, Hayakawa I, Uemura D (1999) Chem Commun 1361
129. Arimoto H, Oishi T, Nishijima M, Kinumi T (2001) Tetra Lett 42:3347
130. Sundram UN, Griffin JH, Nicas TI (1996) J Am Chem Soc 118:13107
131. Klein L, Klugman M, Nave K, Tuohy V, Kyewski B (2000) Nature Med 6:56
132. Firestein GS (2003) Nature 423:356
133. Compton DA (1998) McAlpine's Multiple Sclerosis. 3rd edn. Churchill Livingstone, New York
134. Goodin DS (2000) Exp Opin Invest Drugs 9:655
135. Hall GL, Compston A, Scolding NJ (1997) Trends Neurosci 20:63
136. Donoghue S, Greenlees C (2000) Exp Opin Invest Drugs 9:167
137. Rudick RA (1999) Arch Neurol 56:1079
138. Sela M (1998) Acta Polym 49:523
139. Fricker J (1998) Lancet 351:1792
140. Teitelbaum D, Arnon R, Sela M (1999) Proc Natl Acad Sci USA 96:3842
141. Johnson KP, Brooks BR, Ford CC, Goodman A, Guarnaccia J et al. (2000) Multiple Sclerosis 6:255
142. Fridkis-Hareli M, Santambrogio L, Stern JNH, Fugger L, Brosnan C, Strominger JL (2002) J Clin Invest 109:1635
143. Strominger JL, Fridkis-Hareli M (2004) US Patent Appl 20040038887 (Harvard Univ)
144. Stern JNH, Illes Z, Reddy J, Keskin DB, Sheu E, Fridkis-Hareli M, Nishimura H, Brosnan CF, Santambroigo L, Kuchroo VK, Strominger JL (2004) Proc Natl Acad Sci USA 101:11743
145. Illes Z, Stern JNH, Reddy J, Waldner H, Mycko MP, Brosnan CF, Ellmerich S, Altmann DM, Santambroigo L, Strominger JL, Kuchroo VK (2004) Proc Natl Acad Sci USA 101:11749

146. Fridkis-Hareli M, Rosloniec EF, Fugger L, Strominger JL (1998) Proc Natl Acad Sci USA 95:12528
147. Clapham JC, Arch JRS, Tadayyon M (2001) Pharmacol Therap 89:81
148. Wofford MR, Hall JE (2004) Curr Pharm Design (2004) 10:3621
149. Shi Y, Burn P (2004) Nature Rev Drug Disc 3:695
150. Halford JCG (2004) Current Drug Targets 5:637
151. Bray GA, Tartaglia LA (2000) Nature 404:672
152. Campfield LA, Smith FJ, Burn P (1998) Science 280:1383
153. Phan CT, Tso P (2001) Front Biosci 6:D299
154. Gargouri Y, Ransac S, Verger R (1997) Biochim Biophys Acta 1344:6
155. Hadvary P, Lengfeld H, Wolfer H (1988) Biochem J 256:357
156. Halpern A (2003) Prog Obes Res 9:1045
157. Ballinger A, Peikin SR (2002) Eur J Pharmacol 440:109
158. Harp JB (1998) J Nutr Biochem 9:516
159. Spanswick D, Lee K (2003) Expert Opin Emerging Drugs 8:217
160. Lowe ME (1994) Gastroenterology 107:1524
161. Jozefiak TH, Mandeville WH, Holmes-Farley SR, Arbeeny C, Huval CC, Sacchiero R, Concagh D, Yang K, Maloney C (2001) Polym Prepr 42(2):98
162. Jozefiak T, Holmes-Farley SR, Mandeville WH, Huval CC, Garigapati VR, Shackett KK, Concagh D (2001) US Patent 6299868 (Genzyme Corp)
163. Mandeville WH, Whitesides GM, Holmes-Farley SR (2001) US Patent 6264937 (Genzyme Corp)
164. Holmes-Farley SR (2003) PCT Int Appl WO 2003002130 (Genzyme Corp)
165. Holmes-Farley SR, Mandeville WH, Dhal PK, Huval CC, Li X, Polomoscanik SC (2003) PCT Int Appl WO 2003002571 (Genzyme Corp)
166. Rodgers GP, Noguchi CT, Schechter AN (1994) Sci Am Sci & Med 1:48
167. Persons DA (2003) Cur Opin Mol Ther 5:508
168. Smith CM, Hebbel RP, Tukey DP, Clawson CC, White JG, Vercellotti GM (1987) Blood 69:1631
169. Patrickios CS, Georgiou TK (2003) Cur Opin Col Inter Sci 8:76
170. Schmolka IR (1977) J Am Oil Chem Soc 54:110
171. Kabanov AV, Batrakova EV, Alakhov VY (2002) J Contr Rel 82:189
172. Gibbs WJ, Hagemann TM (2004) Ann Pharmacother 38:320
173. Orringer EP et al. (2001) JAMA 286:2099

Domino Dendrimers

Roey J. Amir · Doron Shabat (✉)

School of Chemistry, Raymond and Beverly Sackler Faculty of Exact Sciences,
Tel-Aviv University, 69978 Tel Aviv, Israel
chdoron@post.tau.ac.il

1	Introduction	60
2	**Design of Domino Dendrons**	67
2.1	Introductory Remarks	67
2.2	Shabat Adaptor Unit	68
2.3	De Groot Adaptor Unit	69
2.4	McGrath Adaptor Unit	70
3	**Examples of Domino Dendrimers**	71
3.1	Geometrically Disassembled Dendrimers	71
3.2	Cascade-Release Dendrimers	72
3.3	Self-Immolative Dendrimers	72
3.4	Kinetic Studies	75
4	**Bioactivation of Domino AB2 Dendritic Prodrugs**	78
4.1	Introductory Remarks	78
4.2	Homodimeric Dendritic Prodrugs	78
4.3	Heterodimeric Dendritic Prodrug	80
5	**Bioactivation of Domino AB3 Dendritic Prodrugs**	81
5.1	Mechanism	81
5.2	Homotrimeric Dendritic Prodrug	83
5.3	In Vitro Advantage of Dentritic Prodrug vs. the Monomeric One	84
5.4	Heterotrimeric Dendritic Prodrug	85
6	**Linearly Disassembled Dendrons**	87
7	**Multi-Triggered Domino Dendrons**	88
7.1	Mechanism	88
7.2	Dendron Structure	88
7.3	Enzymatic Activation of a Multi-Triggered Dendron	89
8	**Outlook**	91
	References	92

Abstract Domino dendrimers have recently been developed and introduced as a potential platform for a single triggered multi-prodrug. Surprisingly, three independent groups reported similar concepts almost simultaneously. These unique structural dendrimers can release all of their tail units, through a domino-like chain fragmentation, which is initiated by a single cleavage at the dendrimer's core. This chapter reviews the recent efforts to design domino-like dendrimers with emphasis on the application of drug delivery. Incorporation of drug molecules as the tail units and an enzyme substrate as the trigger, can generate a multi-prodrug unit that is activated with a single enzymatic cleavage. Dendritic prodrugs, activated through a single catalytic reaction by a specific enzyme, could present significant advantages in the inhibition of tumor growth, especially if the targeted or secreted enzyme exists at relatively low levels in the malignant tissue. Domino dendrimers may also be applied as a general platform for biosensor molecules, used to detect enzymatic activity.

Keywords Biosensor · Cancer · Dendrimer · Prodrug · Self-immolative

Abbreviations
AB2   three arm star dendritic subunit
AB3   four arm star dendritic subunit
ALL   acute lymphoblastic leukemia
AML   acute myeloid leukemia
Boc   *t*-butoxycarbonyl
CPT   campthotecin
DOX   doxorubicin
HEL   human erythroleukemia cell line
HL-60 human acute myeloid leukemia cell line
IC_{50} inhibition concentration (at 50%)
MOLT-3 human T-lineage acute lymphoblastic leukemia cell line
PBS   phosphate buffered saline
PGA   penicillin G amidase
mCPT   monomeric CPT dendritic prodrug
tCPT   trimeric CPT dendritic prodrug
TFA   trifluoroacetic acid

1
Introduction

Dendrimers are treelike molecules with a continually growing impact on chemistry and biology [1, 2]. They are perfectly cascade-branched, highly defined units, characterized by a combination of high-group functionalities and a compact molecular structure. The concept of repetitive growth with branching creates a unique spherical mono-disperse dendrimer formation, which is defined by a precise generation number. The structural precision of dendrimers has motivated numerous studies aimed at biological applications [3], such as, the amplification of molecular effects or the creation of high concentrations of drugs [4], molecular labels, or probe moieties [5].

Among the many different architectures used for dendrimer construction, poly(amidoamine) (PAMAM) dendrimers are the most extensively used and characterized family [6]. PAMAM dendrimers are synthesized by the divergent approach. Starting from a multi-amine core such as ethylenediamine (core multiplicity = 4), the branches of the dendrimer are built by a repetitive sequence of two reactions: (a) Michael addition of a primary amine to two molecules of methyl acrylate; and (b) amidation of the ester by ethylenediamine (Scheme 1). Their ease of synthesis, periphery group modification, and commercialization have led to extensive investigation of the biological properties of PAMAM dendrimers of various generations. Comprehensive reviews have been published in the literature covering PAMAM and other families of dendrimers and their biological applications [7–10]. In this introduction we review the most recent reports of dendrimers in the field of medicinal chemistry. Dendrimers can serve as a drug delivery platform either by carrying the drug load as the outer shell or by encapsulation of the drug molecules in the dendrimer cavities.

D'Emanuele et al. used G3 and lauroyl-G3 PAMAM dendrimers to form prodrugs of propranolol (Scheme 2), which is an example of a P-glycoprotein (P-gp) substrate with poor water solubility [11]. The PAMAM propranolol conjugates were found to have higher permeability through cell monolayers and reduced P-gp drug elimination in comparison with free propranolol. This

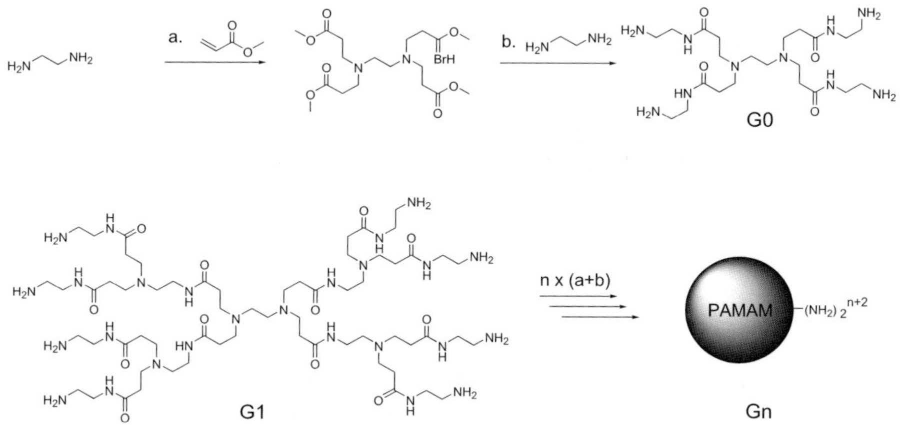

Scheme 1 PAMAM synthesis and structures

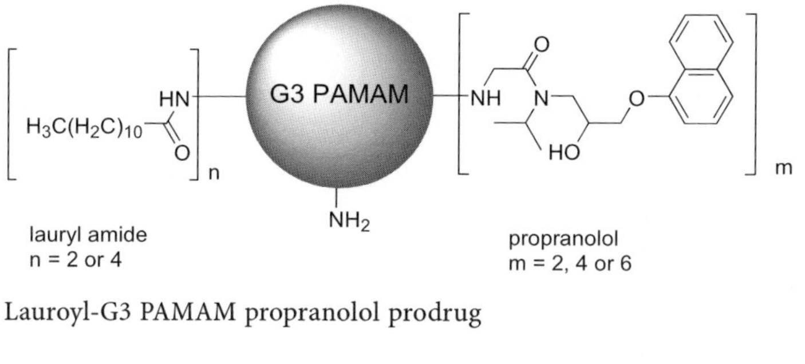

Scheme 2 Lauroyl-G3 PAMAM propranolol prodrug

use of dendrimers may be of great importance in reducing the effects of the intestinal P-gp on drug absorption, thus improving the oral bioavailability of low solubility drugs that are P-gp substrates.

Lin and coworkers recently described a gene delivery device based on a mesoporous silica nanosphere as a dock for multiple covalently-attached G2 PAMAMs [12]. The dendritic subunits of the nanospheric delivery device (Fig. 1) were used to bind plasmid DNA that encodes an enhanced green fluorescent protein. The silica nanosphere PAMAM conjugates were evaluated as a potential transmembrane gene delivery system. Their stability, transfection efficacy and mammalian cell membrane permeability were investigated with different cell lines. A different architecture, used by Park and coworkers, applied poly(ethylene glycol) (PEG) and PAMAM to synthesize a triblock copolymer [13]. Commercial PEG amine was used as a core and two PAMAM dendrons were synthesized through the divergent method. The PAMAM-PEG-PAMAM (Scheme 3) was shown to have higher water solubility and lower cytotoxicity than with PAMAM dendrimers of a similar size. In

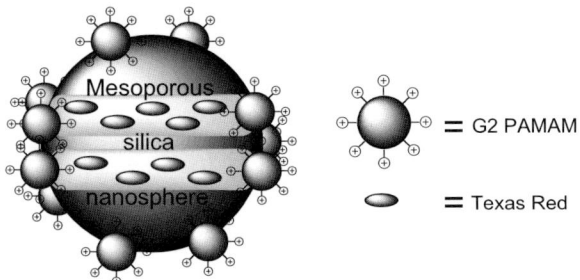

Fig. 1 A PAMAM-capped mesoporous silica nanosphere-based gene delivery device

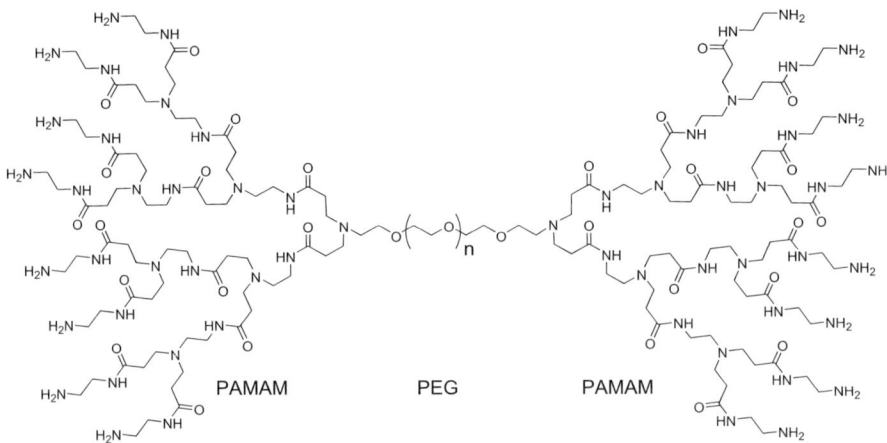

Scheme 3 PAMAM-PEG-PAMAM triblock copolymer

addition the triblock copolymer achieved high transfection efficiency compared with other gene delivery carriers such as polyethyleneimine (PEI) [14]. Park's work demonstrates the modularity of dendritic devices as it combines the affinity of cationic PAMAM dendrons with negatively charged plasmid DNA and the utility of PEG to improve the water solubility and biocompatibility of its complexes.

In order to achieve site-specific drug targeting, a conjugate of an antibody, as a targeting moiety, and a PAMAM dendrimer (Scheme 4) was designed by Baker and [15, 16]. The PAMAM dendrimer was used to covalently bind fluorophores as drug models. Confocal microscopy experiments showed enhanced binding and internalization of the antibody-dendrimer conjugates over that of the dendrimer alone. A new dendritic architecture, based on a different poly(amidoamine), was recently reported by Gust and Schluter [17]. G1 and G2 dendrimers (Scheme 5) were synthesized and their in vitro toxicity and cellular uptake were investigated. Internalization and intracellular distribution were studied using fluorescent-labeled dendrimers in confocal microscopy experiments. Gust and Schluter results showed that the interior of a dendrimer has little to do with its cytotoxicity. Although, it is commonly thought that the interior of low-generation dendrimers is accessible to the surrounding.

Park and coworkers recently reported the synthesis of polyglycerol dendrimers (PGDs) and their application as an alternative method of hydrotropic solubilization of poorly soluble drugs [18]. PGDs of the 4th and 5th generations (Scheme 6) were shown to significantly enhance the water solubility of paclitaxel, a commonly used anticancer drug with low water solubility, over that of PEG 400 (commonly used as a co-solvent or a hydrotropic agent). Paclitaxel solubility was found to be dependent on the dendrimer generation and NMR spectra suggested that the aromatic rings and some methylene groups of the drug were surrounded by PGDs.

Niidome and coworkers studied a G6 dendritic poly(L-lysine) (K6G) as a non-viral gene delivery device [19]. L-lysine has two primary amine moieties that serve to link with two other lysines through amide bonds, thus cre-

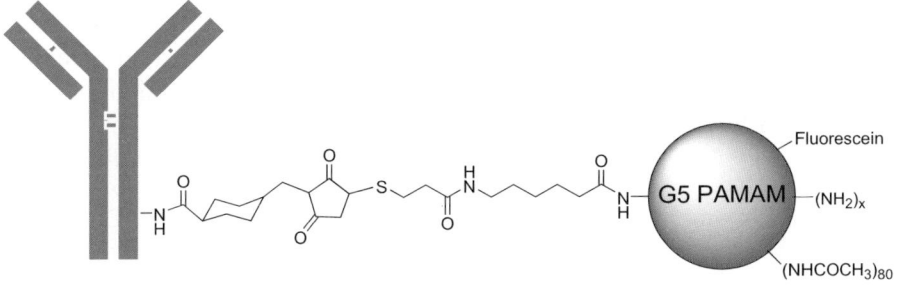

Scheme 4 Antibody-G5 PAMAM dendrimer conjugate

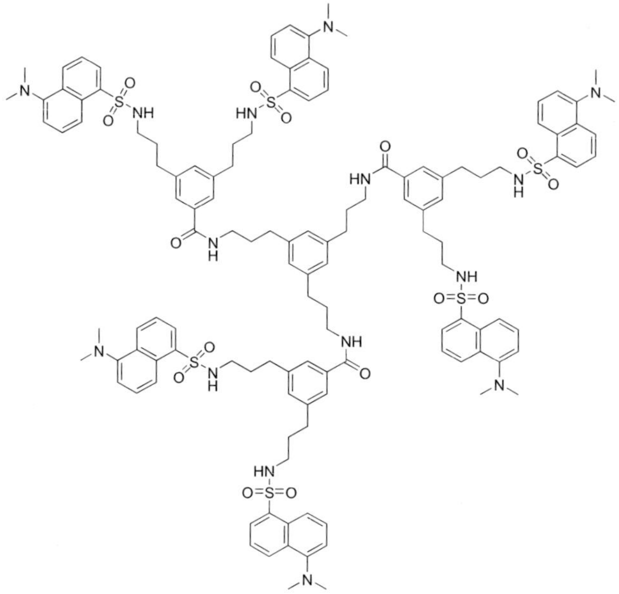

Scheme 5 Structure of Gust and Schluter's poly(amidoamine) G2 dendrimer

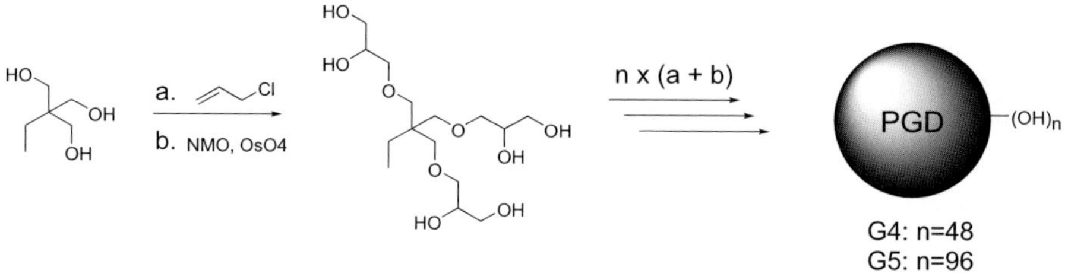

Scheme 6 Preparation of polyglycerol dendrimers

ating a dendritic structure (Scheme 7). The dendrimer-DNA complex biodistribution was investigated in vivo and compared with gene carriers based on liposomes or PEI dendrimers. Although KG6 had a longer circulation time in the blood and was accumulated through the EPR effect without the need to add additional components, such as PEG, no significant gene expression was observed. Possible explanations for the low gene expression could be either that the DNA-K6G complex could not be internalized or that K6G binds DNA too strongly to be released.

Diaminobutane poly(propylene imine) dendrimers (DAB) were studied by Paleos et al. as a drug delivery system [20]. A dendrimer of the 5th generation with 64 amino end-groups (DAB-64) was decorated with PEG chains and guanidinium moieties (Scheme 8) in order to induce protective and targeting properties, respectively. The release of encapsulated molecules, such as pyrene, was triggered by titration with hydrochloric acid followed by the

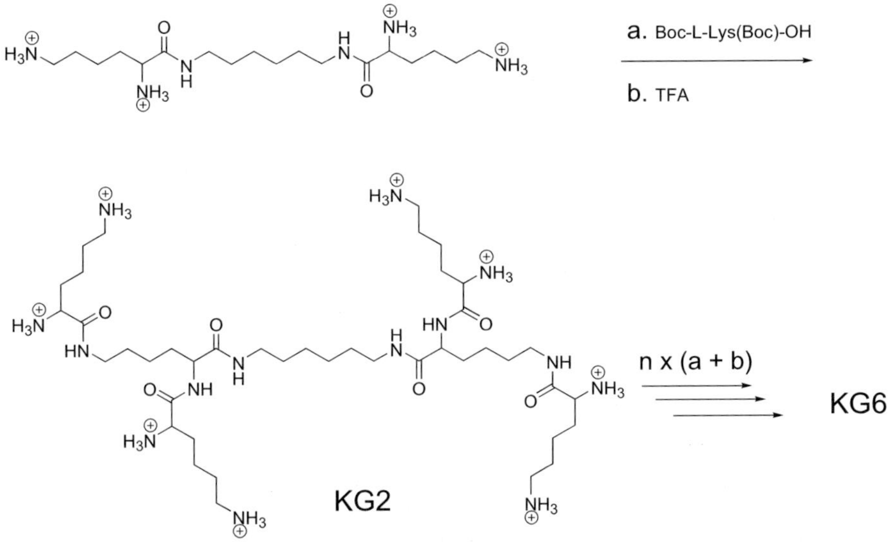

Scheme 7 Synthesis of dendritic poly(L-lysine)

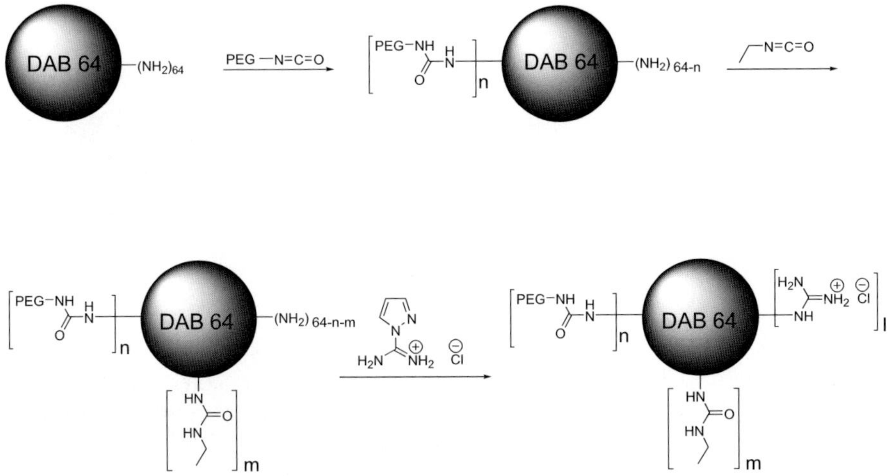

Scheme 8 Acid and salt-triggered DAB drug delivery device

addition of a sodium chloride solution. Upon acidification, the dendrimers' cavities become hydrophilic due to the protonation of the amines in the dendritic structure and the encapsulated hydrophobic molecules migrated to the PEG periphery. Addition of sodium chloride resulted in cationization of the PEG moieties through complexation of the alkali ions, thus, releasing the encapsulated compounds.

The increasing recognition of the importance of polyvalent receptor-ligand interactions between carbohydrates and proteins in many aspects of cell surface-mediated immuno-regulation, have motivated the exploit of dendrimers as platforms for polyvalent interactions [21]. Recently, Shaunak and colleagues synthesized PAMAM dendrimers with D(+)-glucosamine and D(+)-glucoseamine 6-sulfate (Scheme 9) and investigated their immunomodulatory and antiangiogenic properties, respectively [22]. Anionic G3.5 PAMAM dendrimers with 64 carboxylic acid end groups were conjugated to the aminosaccharide to yield the desired compounds with 14% loading of the amino carbohydrates (e.g. nine saccharide molecules per dendrimer). The combination of both aminosaccharide dendritic conjugates resulted in an efficient and synergetic increase in the long-term success of eye surgery from 30% to 80% in a rabbit model, due to the prevention of scar tissue formation.

Dendrimers may serve other functions in drug delivery devices than their common use as drug/gene carriers. Fernandez and colleagues conjugated dendritic structures to β-Cyclodextrin (β-CD) to create a complex nano-device (Scheme 10) for drug targeting [23]. β-CD was chosen to serve

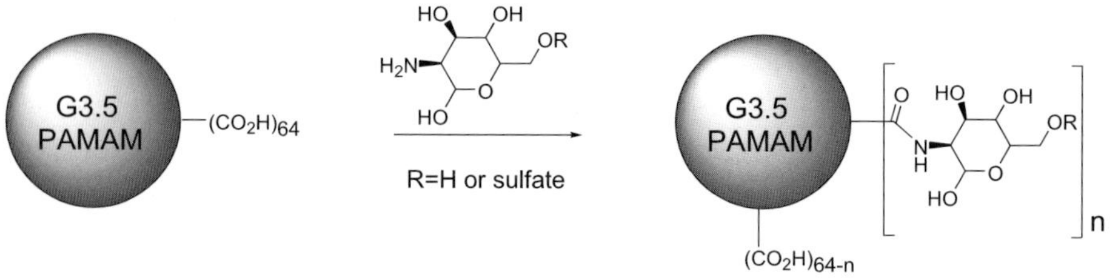

Scheme 9 Polyvalent dendrimer glucosamine conjugates

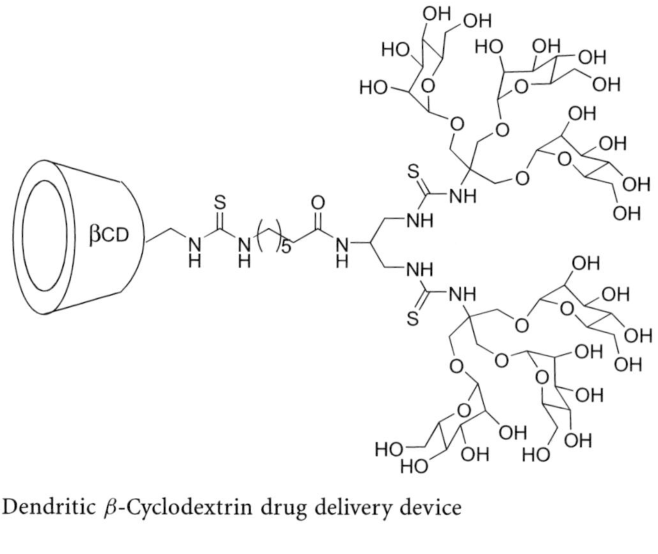

Scheme 10 Dendritic β-Cyclodextrin drug delivery device

as a drug container, due to its ability to encapsulate hydrophobic guest molecules of the appropriate size, such as the anti-cancer drug paclitaxel. The dendrimer was used as a platform for the targeting moieties; saccharide markers were covalently attached as end groups enabling the complex to bind to carbohydrate-binding protein (lectin) targets. High drug solubilization capacity and lectin affinity were proved for this modular delivery device.

Since the first dendrimers were published in the early 1980s, thousands of works describing dendritic structures and applications have been published. Furthermore, dendrimers are accounted as one of four major classes of macromolecular architectures [1]. The diverse examples of dendrimers as drug/gene delivery devices clearly express the broad use of dendrimers and their potential in the quest for efficient and selective drug delivery systems. Dendrimers' capability to covalently bind or encapsulate multiple molecular components to form nanostructures has been the major utility in designing dendritic applications and devices [24]. These applications rely mainly on the high-group functionality rather than their unique structural perfection. The structural properties of a dendrimer could potentially be harnessed for amplifying a signal generated at the focal point towards the high-group functionalities. This review is focused on the recent concept of disassembled domino-like dendrimers. Recently, three groups, almost simultaneously, reported a novel concept describing the design and synthesis of dendritic structures with a trigger that initiates the fragmentation of the dendrimer molecule into its building blocks in a self-immolative manner with the consequent release of the tail-high-group units. All three exploit the fact that the dendrimer skeleton can be constructed in such a way that it can be made to disintegrate into known molecular fragments once the disintegration process has been initiated. Various terminologies were used: "self-immolative dendrimers" by our group [25, 26], "cascade-release dendrimers" by de Groot [27] and "geometrically disassembled dendrimers" by McGrath [28, 29]. The fragmentation of dendrons is similar to sequential dominos falling onto each other.

2
Design of Domino Dendrons

2.1
Introductory Remarks

The design of a first generation domino dendron is based on a molecular adaptor with three functional groups. Two identical functionalities are linked to reporter molecules and the third is attached to a trigger (Fig. 2 I). The cleavage of the trigger, initiates a self-immolative reaction sequence that

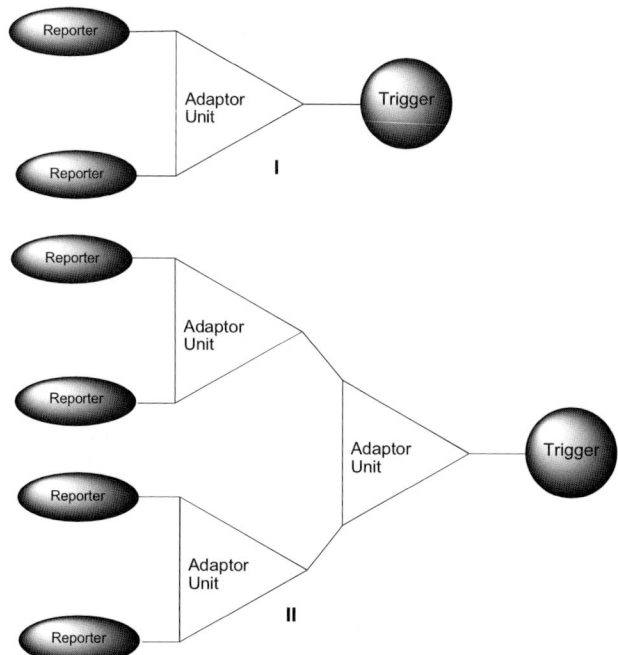

Fig. 2 I Graphical structure of a first generation domino dendron with a trigger and two tail units. II Graphical structure of a second generation domino dendron with a trigger and four tail units

leads to a spontaneous release of the two reporter molecules. The adaptor molecule can be linked to two additional identical units, each attached to two reporter molecules (Fig. 2 II). The head position of the first adaptor unit is linked to a trigger. In this approach, a second-generation dendron can be prepared and similarly, the design can be extended to higher generations of dendrons. The cleavage of the trigger will initiate self-immolative chain reactions that will consequently fragment the dendrimer and release all of the tail molecules.

The ability to synthesize such a domino dendrimer relies mainly on finding a molecule with the structural properties described for the adaptor unit. Three independent groups, including ours, suggested different yet related molecules as the molecular adaptor for the dendrimer building block.

2.2
Shabat Adaptor Unit

Our dendrimer's adaptor unit is based on 2,6-bishydroxymethyl-*p*-cresol **7**, a commercially available compound with three functional groups (Scheme 11).

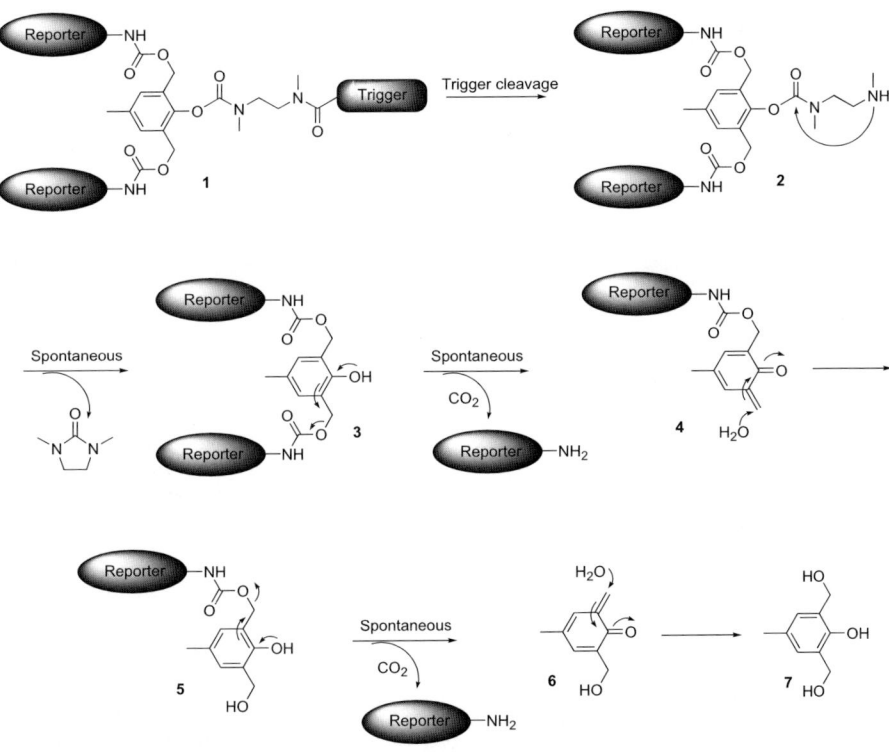

Scheme 11 Proposed disassembly pathway of 2,6-bis(hydroxymethyl)-*p*-cresol-based dendrons

The two hydroxybenzyls are attached through a carbamate linkage to reporter molecules and the phenol functionality is linked to a trigger through a short spacer N,N'-dimethyl-ethylenediamine (**1**). The cleavage of the trigger initiates a self-immolative reaction sequence of amine intermediate **2**, starting with spontaneous cyclization, to form an N,N'-urea derivative. The generated phenol **3** undergoes a 1,4-quinone-methide rearrangement, followed by spontaneous decarboxylation to liberate one of the reporter molecules. The quinone-methide species **4** is rapidly trapped by a water molecule (from the reaction solvent) to form a phenol (**5**), which again undergoes a 1,4-quinone-methide rearrangement to liberate the second reporter molecule. The generated quinone-methide species **6** is again trapped by a water molecule to form 2,6-bishydroxymethyl-*p*-cresol **7**.

2.3
De Groot Adaptor Unit

De Groot and coworkers chose to use a different adaptor molecule, described previously by Firestone [30], as the dendrimer's building block. The

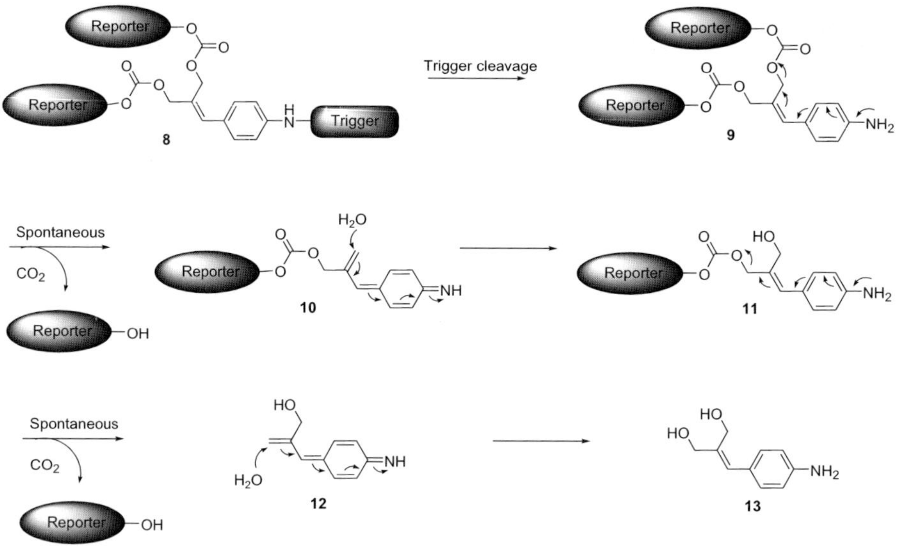

Scheme 12 Proposed disassembly pathway of 2-(4-aminobenzylidene) propane-1,3-diol-based dendrons

double-release self-elimination AB2-type monomer 2-(4-aminobenzylidene) propane-1,3-diol (**13**) and the proposed mechanism of the release of two leaving groups upon activation are depicted in Scheme 12. The multiple-release system is stable as long as the amine function in **8** is capped by a protecting group. Unmasking the protected amine **8** triggers two 1,8-elimination reactions from amine **9**, in which two molecules of CO_2 and two reporter groups are liberated. The intermediate non-aromatic species **10** that is formed after the first self-elimination is trapped by a nucleophile, such as water, to regenerate an aromatic species **11** that can undergo the second self-elimination. Quenching of the intermediate with water will generate an aminodiol building block **13**.

2.4
McGrath Adaptor Unit

McGrath and coworkers used 2,4-bis(hydroxymethyl)phenol **19** as the adaptor unit for their dendrimer (Scheme 13). Removal of the trigger group from the 2,4-bis-(hydroxymethyl)phenol-based dendrimer subunit **14** results in a 1,6-elimination and the formation of the quinone-methide (**16**). The latter is trapped by an appropriate nucleophile under the reaction conditions, consistent with the electrophilic nature of quinone methides. The resulting phenol **17** (or phenoxide under basic conditions) undergoes 1,4-elimination to liberate a second equivalent of alkoxide and o-quinone methide **18** which, in turn is trapped by a nucleophile to yield the fully cleaved phenol **19**.

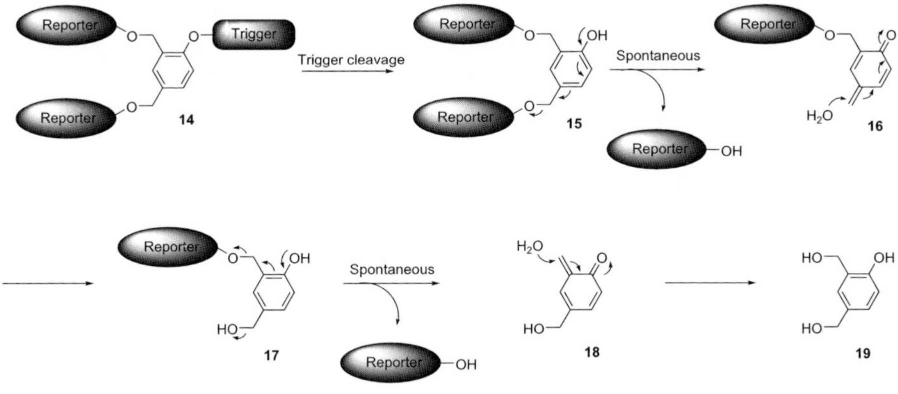

Scheme 13 Proposed disassembly pathway of 2,4-bis(hydroxymethyl)phenol-based dendrons

3
Examples of Domino Dendrimers

3.1
Geometrically Disassembled Dendrimers

Both McGrath and de Groot succeeded in synthesizing the first and second generation of their domino dendrimers with different triggers and reporters. The chemical structure of the McGrath dendrons is shown in Scheme 14. Dendrons **20** and **21** consist of a single allyloxy residue at the focal point and 2,4-branched benzyl ether dendrimer subunits. When deprotection of the allyl group occurs as the initial triggering event, subsequent cleavages should result in a disassembly, according to Scheme 13, toward the dendron periphery. *p*-nitrophenoxy moieties were intentionally installed at the periphery

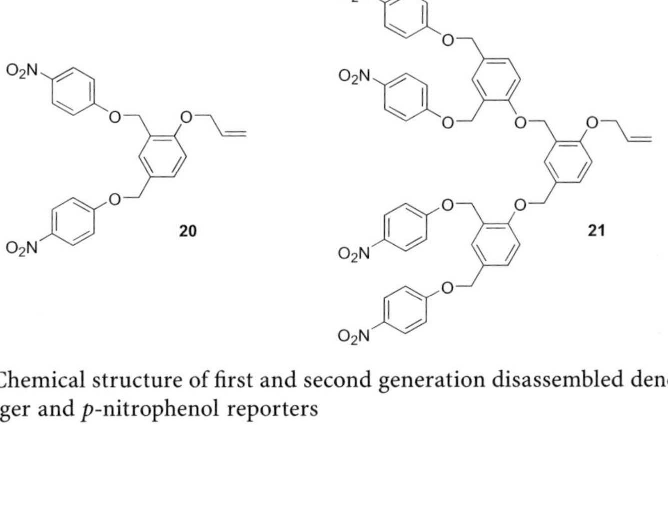

Scheme 14 Chemical structure of first and second generation disassembled dendrons with an allyl trigger and *p*-nitrophenol reporters

of each dendron so that complete cleavage would be indicated by the UV absorbance of liberated *p*-nitrophenoxide ions. To investigate the geometric disassembly process, compounds **20** and **21** were subjected to typical allyl deprotection conditions. Once begun, *p*-nitrophenoxide generation from **20** and **21** was complete within minutes. The final absorbance values observed at 431 nm indicated disassembly of dendrons **20** and **21** in yields of ca. 95% based on the measured absorptivity of *p*-nitrophenoxide under the reaction conditions.

3.2
Cascade-Release Dendrimers

De Groot's dendrons (Scheme 15) were constructed with a nitro functionality as the trigger (activation is achieved upon reduction of the nitro group to amine) and the anticancer drug taxol as the reporter units. The nitro function in dendrons **22** and **23** was then reduced under mild conditions (Zn, acetic acid). Analysis by thin-layer chromatography indicated complete disappearance of the starting material and formation of free taxol. ^{1}H NMR spectroscopic studies also confirmed the complete release of the taxol molecules. However, there is no data yet to prove that taxol dendrons can be activated under physiological conditions.

3.3
Self-Immolative Dendrimers

Our group was able to proceed one step further and prepare domino-like dendrons up to a third generation (Scheme 16). Initially we used simple reporter groups like aminomethyl-pyrene and 4-nitroaniline. The trigger was constructed from photo-cleavable or acid-cleavable protecting groups. To prove the self-immolative mechanism, we synthesized the first generation dendron (**24**). The compound was dissolved in methanol and the solution radiated with UV-light ($\lambda = 360$ nm) to cleave the trigger and 10% triethylamine was added to initiate the self-immolative reactions (the triethylamine is needed to generate the mild basic media which is needed for the quinonemethide rearrangement). The release of aminomethyl-pyrene **29** was monitored by HPLC and the results are shown in Fig. 3. The cleavage of the photo-labile trigger generated amine **28**, which gradually degraded to the tail-units through the previously explained self-immolative process. The release of aminomethyl-pyrene was complete after 11 h. Since no intermediates other than amine **28** were observed, we concluded that the rate limiting step of the self-immolative sequence is the cyclization of amine **28** to form a N,N'-dimethyl-urea derivative and a phenol which is rapidly rearranged to release the tail-units. We characterized amine **28** by HRMS-analysis and HPLC comparison with a reference compound.

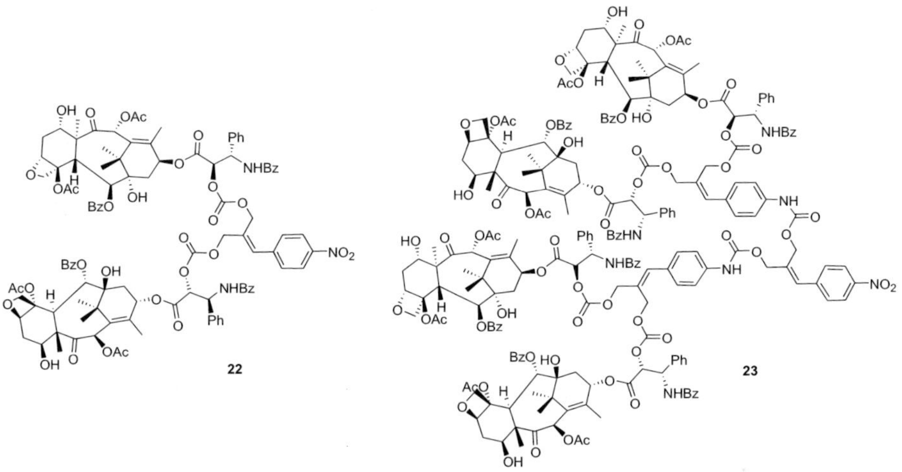

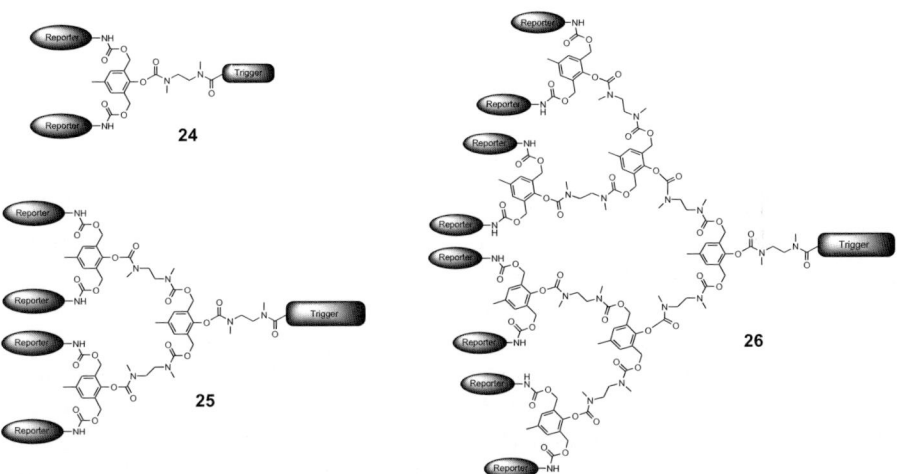

Scheme 15 Chemical structure of first and second generation disassembled dendrons with a nitro trigger and taxol reporters

Scheme 16 Chemical structure of first, second, and third generation disassembled dendrons with a general trigger and reporters

Next, we synthesized a second generation of self-immolative dendron with a similar tail-unit and photo-labile trigger (like compound **25**). We repeated the previous experiment and monitored the release of the tail-molecules. It was clearly observed (data not shown) that upon cleavage of the trigger, the self-immolative release of aminomethyl-pyrene is initiated and completed after 21 h.

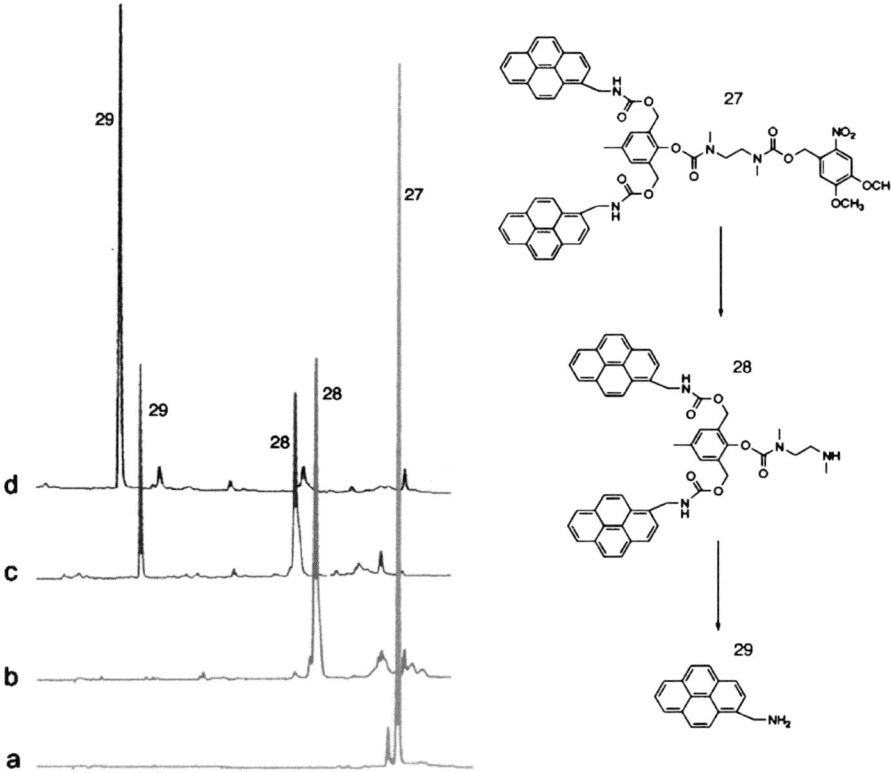

Fig. 3 HPLC chromatograms of first generation self-immolative dendron activation to release aminomethyl-pyrene (compound **27** [50 μM] in methanol with 10% triethylamine). **a** Before radiation. **b** After radiation, $t = 0$. **c** $t = 4$ h. **d** $t = 11$ h

We were also successful in achieving the synthesis of a third-generation dendron with 4-nitroaniline reporter groups (Scheme 17). 4-Nitro-aniline is easily observed when unconjugated due to its yellow color. Dendron **30** was prepared with a trigger group (BOC) that can be chemically removed by trifluoroacetic acid to form amine **31**. The deprotected dendron **31** was dissolved in methanol with 10% triethylamine and the release of 4-nitro-aniline was monitored by HPLC and UV analysis. The expected pattern of the self-immolative process was observed (Scheme 17). The intermediates **32** and **33** were gradually generated and disappeared to finally release eight molecules of 4-nitro-aniline. The dendrimers were found to be highly stable in control experiments, as long as the trigger was not removed, and no decomposition was observed for at least 72 h.

3.4
Kinetic Studies

The dendrimers' platform is designed to disassemble upon triggering through a process of self-immolative chain fragmentation, based on cycli-

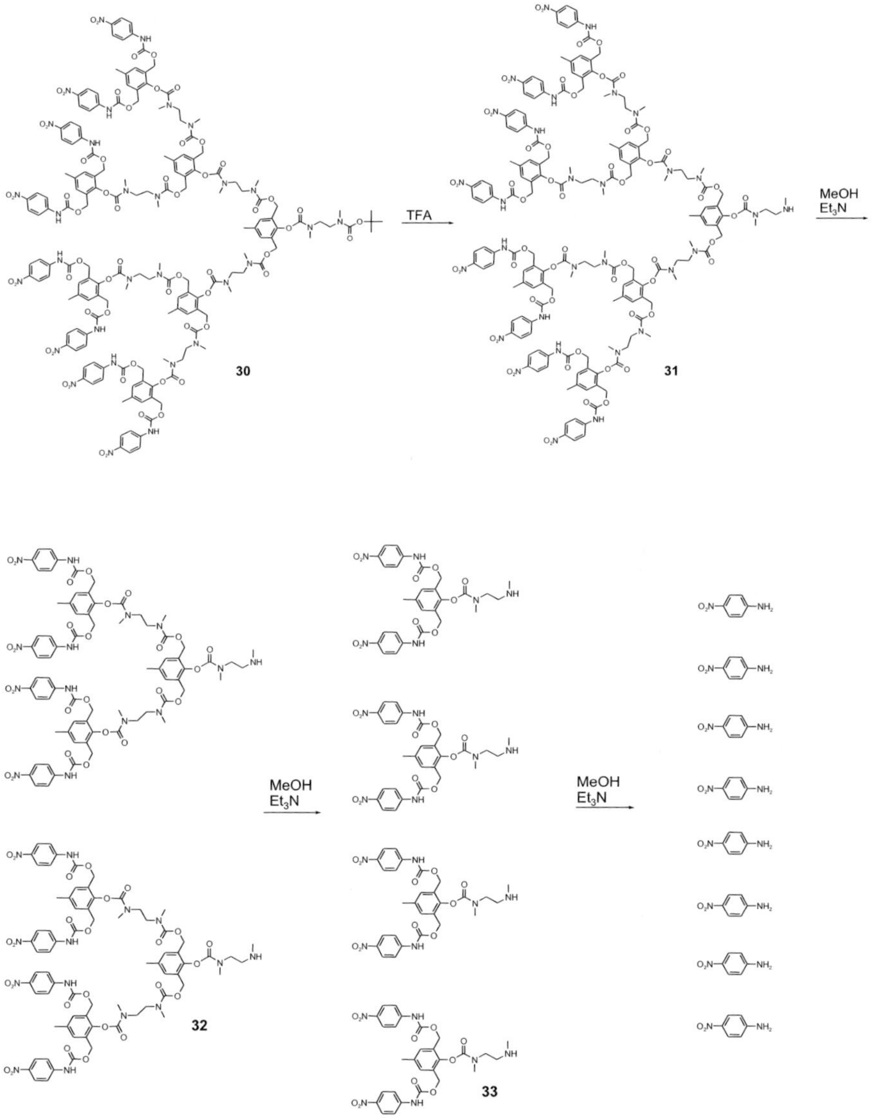

Scheme 17 Third-generation self-immolative dendron triggered with TFA to release eight tail units of 4-nitroaniline

zation and elimination reactions. An additional support for the suggested self-immolative release mechanism emerges from further analysis of the kinetic data, collected by HPLC experiments [31]. The cyclization reaction of each dendritic amine intermediate (compound type 2) was found to be the rate limiting steps during the dendron disassembly (first order reaction). Therefore, the cyclization rate constant (k_n) can be calculated from a plot of the natural logarithm of the dendron concentration (n is the generation index) as a function of time. Excellent correlation was found for all three generations (Figs. 4 and 5) and the rate constants k_3, k_2 and k_1 were all found to be identical ($k_{1-3} = 2.2 \times 10^{-3}$ min^{-1}). Since the rate-determining step of the domino reactions is the cyclization (forming N,N'-dimethyl urea derivative), it is obvious that the fragmentation rate constants should be similar for G1, G2, or G3 (G1: first-generation, G2: second-generation, G3: third-generation).

The equations for the release of the G1 amine-intermediate 24 from the G2 amine intermediate 25, are described by

$$\frac{\partial}{\partial t}[G2] = k_2[G2]_{(t)} \tag{1}$$

$$\frac{\partial}{\partial t}[G1] = 2k_2[G2]_{(t)} - k_1[G1]_{(t)} \tag{2}$$

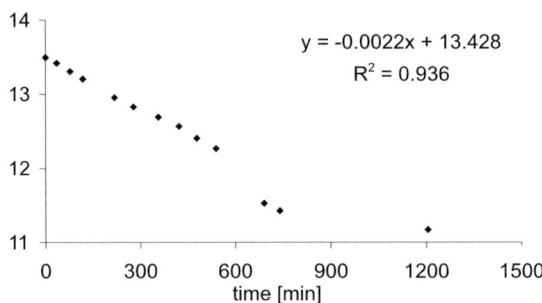

Fig. 4 G1-Dendron disassembly: $Ln[G1]_{(t)}$ as a function of time

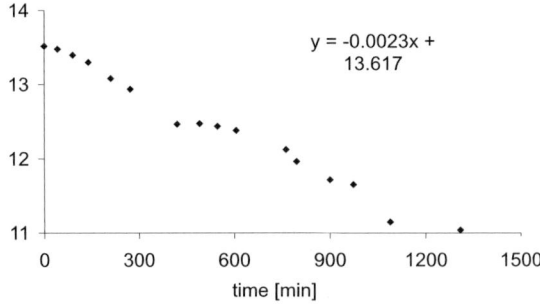

Fig. 5 G2-Dendron disassembly: $Ln[G2]_{(t)}$ as a function of time

The solution of Eq. 1 is given by

$$[G2] = [G2]_{(t=0)} e^{-k_2 t} \qquad (3)$$

We were able to solve Eq. 2 and to predict the time dependence of a G1 amine intermediate concentration $[G1]_{(t)}$ during the fragmentation of a G2 dendron (Eq. 4), based on Eq. 3 and the similar values of k_1, k_2 and k_3.

$$[G1] = 2k[G2]_{(t=0)} t e^{-kt} \qquad (4)$$

The theoretical results show good correlation with the experimental results regarding curve shape and maximal value, as shown in Fig. 6a. Finally, the time dependence of the reporter concentration $[R]$ was obtained by solving the following rate equation:

$$\frac{\partial}{\partial t}[R] = 2k[G1]_{(t)}, \qquad (5)$$

is given by

$$[R]_{(t)} = 4[G2]_{(0)}(1 - e^{-k_+ t} - k_+ t e^{-k_+ t}). \qquad (6)$$

An excellent correlation was obtained between the calculated and experimental $[R]$ values (Fig. 6b), which supports the assumption that a single rate constant is suitable for describing the dissociation process.

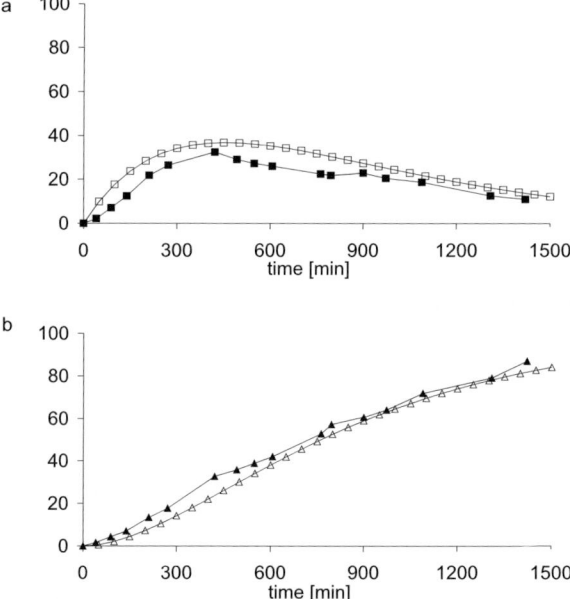

Fig. 6 **a** Formation of intermediate **28**: empiric (-■-), theoretical calculation (-□-). **b** Formation of the reporter from **28**: empiric (-▲-) and theoretical calculation (-△-)

Figures 6a and b support our mechanistic and kinetic characterization of the domino dendron system. The kinetic analysis of G1 and G2 dendrons may serve as a reliable method to characterize other domino dendritic systems. It can also allow a better understanding and evaluation of the kinetic contributions of introducing different substituents on the core benzene ring, and of modifying the linker component of the triggering substrate.

4
Bioactivation of Domino AB2 Dendritic Prodrugs

4.1
Introductory Remarks

Several anti-cancer prodrugs have been designed for selective activation in malignant tissues by a specific enzyme, which is targeted or secreted near tumor cells [32]. The release of the free drug by a specific enzyme only takes place upon cleavage of a prodrug protecting group. The circumstances under which a cleavage event will release one molecule of free drug may limit the total amount of the targeted drug, depending on the rate and concentration of the specific enzyme [33]. Single triggered disassembled dendrimers could potentially introduce significant advantages for the prodrug approach. Incorporation of drug molecules as the tail units and an enzyme substrate as the trigger, can generate a multi-prodrug unit that will be activated upon a single enzymatic cleavage. Domino dendritic prodrugs may open up new opportunities for targeted drug delivery. In contrast with conventional dendrimers, domino dendrimers are fully degradable and can be excreted easily from the body. The cleavage effect of a tumor-associated enzyme or a targeted one, can be amplified and therefore, may increase the number of active drug molecules in targeted tumor tissues.

4.2
Homodimeric Dendritic Prodrugs

Two examples of a dimeric-prodrug were synthesized with the anti-cancer drug doxorubicin and camptothecin [34]. Catalytic antibody 38C2 [35] was used as the activating enzyme. Antibody 38C2 catalyzes a sequence of *retro*-aldol *retro*-Michael cleavage reactions, using substrates that are not recognized by human enzymes [36]. Therefore, non-specific prodrug activation should be minimal. Furthermore, the antibody has demonstrated efficacy in activating several prodrugs in vitro and in vivo [37]. A dramatic 75% decrease in subcutaneous (s.c.) tumor size has been observed in mice that received a combination of intratumoral injections of antibody 38C2 and systemic treatments with an etoposide prodrug [38]. The *retro*-aldol *retro*-

Michael substrate of antibody 38C2 was attached to the adaptor platform through a self-immolative linker of adequate length, to avoid steric hindrances accompanying the complex structure of the doxorubicin (DOX) molecule (Scheme 18) or the campthotecin (CPT) (Scheme 19).

The activation of dimeric-prodrugs **34** and **35** was compared with monomeric-prodrugs **34a** and **35a**, using a cell-growth inhibition assay of the Molt-3 cell line. The IC$_{50}$s of the monomeric and dimeric-prodrugs were found to be almost identical, and between 50- and 200-fold less toxic than the free drugs. When catalytic antibody 38C2 was added, all prodrugs were

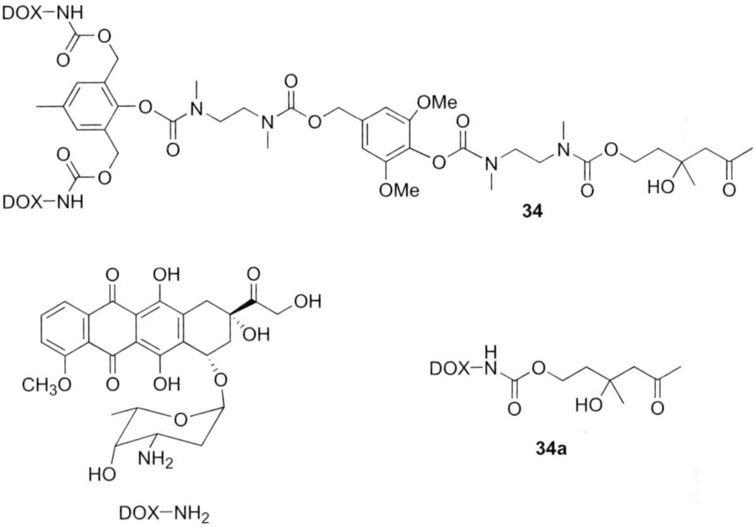

Scheme 18 Chemical structure of prodrugs **34** and **34a**

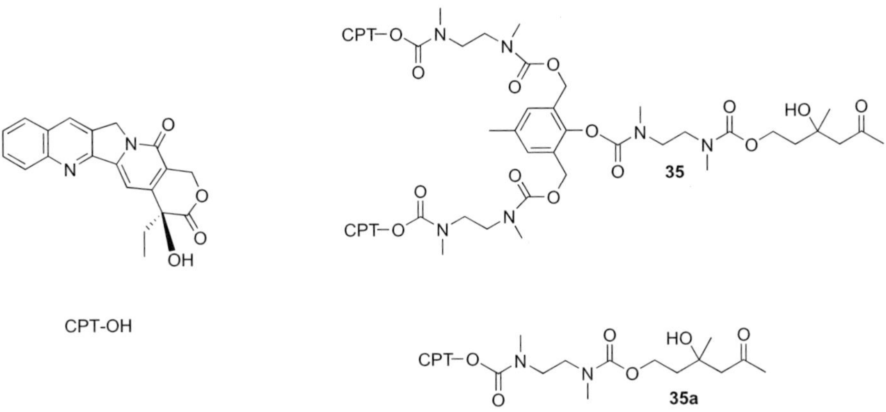

Scheme 19 Chemical structure of prodrugs **35** and **35a**

activated. However, while the activity of the monomeric-prodrug had shifted to a 10-fold difference from that of its parent drug, the dimeric-prodrug was shown to be about four times more active upon addition of 38C2, meaning that more toxicity was achieved using the dimeric-prodrug and 38C2 in comparison with the monomeric-prodrug and the same concentration of antibody. Importantly, the toxicities of the monomeric-prodrug and the dimeric-prodrug remain in a similar range, indicating the relative stability for hydrolysis of the drug molecules' linkages with the platform.

4.3
Heterodimeric Dendritic Prodrug

With these results in hand, we were motivated to synthesize a heterodimeric prodrug, constructed of a combination of DOX and CPT. Heterodimeric prodrug **36** was prepared with a retro-aldol retro-Michael trigger as a substrate for antibody 38C2 [34].

The bioactivation of dimeric-prodrug **36** was compared with a 1 : 1 combination of monomeric-prodrugs **34a** and **35a** at the appropriate concentrations. The in vitro results using a cell-growth inhibition assay of the Molt-3 cell line are shown in Fig. 7. The IC_{50}s of both prodrug **36** and the combination **34a+35a** are almost the same. However, in the presence of antibody 38C2 the IC_{50} of prodrugs **34a+35a** shifted to only 8 nM while the IC_{50} of prodrug **36** decreased to 0.17 nM. This effect is remarkable and shows a clear advantage of the heterodimeric prodrug over the combination of two monomeric prodrugs. The prodrug bioactivation is much more efficient if two drug molecules are attached to a single common masking enzymatic substrate rather than two separated substrates. Importantly, no drug release was observed when prodrug **36** was incubated in a cells' extract for 24 h.

The best results for the dendritic compounds were obtained with a heterodimeric prodrug. Toxicity for prodrug **36**, constructed of DOX and CPT, was about 50-fold higher than the activity measured using a combination of two monomeric prodrugs (**34a** and **35a**) when bioactivation was performed.

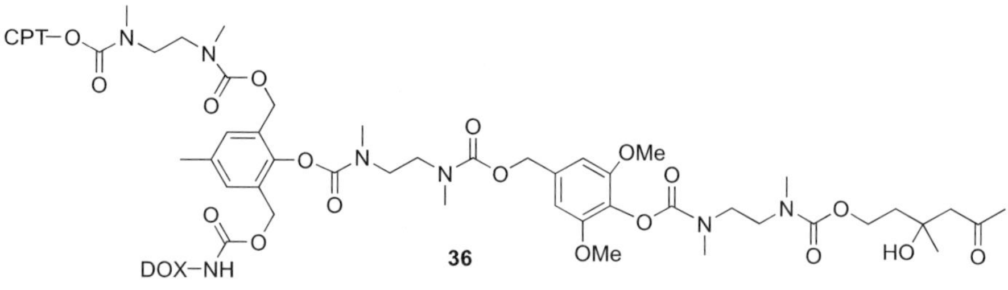

Scheme 20 Chemical structure of heterodimeric-prodrug **36**, containing an enzymatic trigger substrate of catalytic antibody 38C2

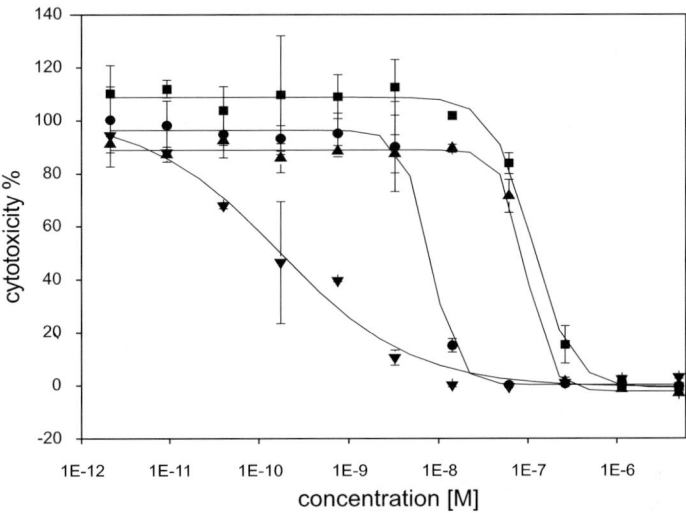

Fig. 7 Growth inhibition assay of human Molt-3 leukemia cell line; cells were incubated for 96 h. (■) pro-CPT **35a** + pro-DOX **34a**, (●) pro-CPT **35a** + pro-DOX **34a** + 1 µM 38C2, (▲) heterodimeric prodrug **36**, (▼) heterodimeric prodrug **36** + 1 µM 38C2

One reason for this effect was discussed above and is merely related to the number of cleaving events required for activation. However, additional explanation is required in order to fully understand the significant effect. At this time, we believe that in the bioactivation of the two monomeric prodrugs (**34a** and **35a**), there may be a case of competitive substrates. Thus, if one prodrug is a better substrate than another, the overall release of the free drugs combination is inhibited in comparison with the activation of the heterodimeric prodrug. Importantly, the toxicity of the heterodimeric prodrug, in the absence of catalytic antibody 38C2, was similar to that measured for the combination of the monomeric prodrugs. Different drugs can be introduced on the dendritic platform in order to achieve synergetic effects and precise drug combinations may be tailored for specific types of cancer.

5
Bioactivation of Domino AB3 Dendritic Prodrugs

5.1
Mechanism

To further explore the advantages of a dendritic prodrug platform we synthesized a new system, based on an AB_3 dendritic unit (Scheme 21), where three drug molecules (B) are linked to one enzymatic substrate (A). The trigger is activated upon enzymatic cleavage and is designed to release the three

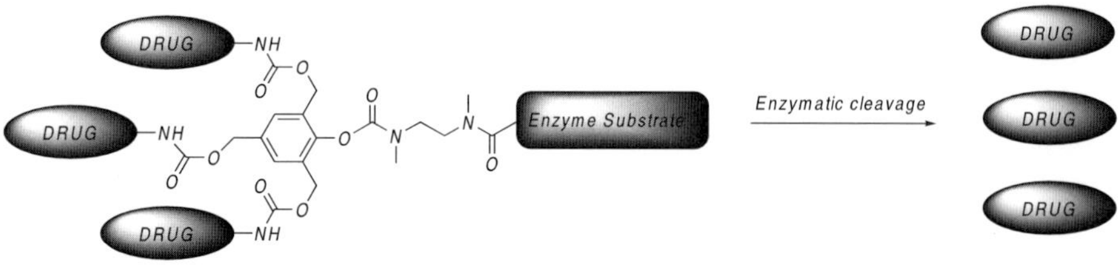

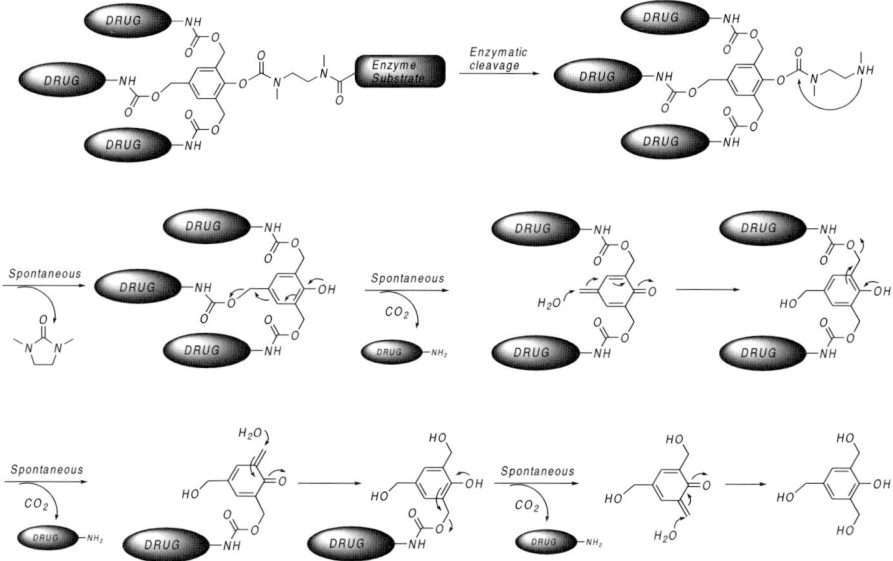

Scheme 21 General structure of a single triggered trimeric prodrug

Scheme 22 Suggested mechanism for drug release from the dendritic platform. The enzymatic cleavage initiates a self-immolative reaction sequence based on an internal cyclization and a triple quinone-methide rearrangement

drug molecules through a mechanism based on self-cyclization and a triple quinone-methide rearrangement [39] (Scheme 22). Although a similar conceptual molecule was previously described by de Groot, it was not examined as a prodrug system [27].

In order to evaluate the self-cyclization and a triple quinone-methide rearrangement, we synthesized a model system constructed from three molecules of *p*-nitroaniline to mimic the drug units and a butoxycarbonyl (Boc) protecting group to mimic the enzymatic substrate. Cleavage of the Boc group should trigger the release of the *p*-nitroaniline moieties as shown in Scheme 23. Initially, we set out to examine whether the release could occur under physiological conditions. The Boc group was removed with dry hydrogen chloride and the product was then incubated in phosphate-buffered

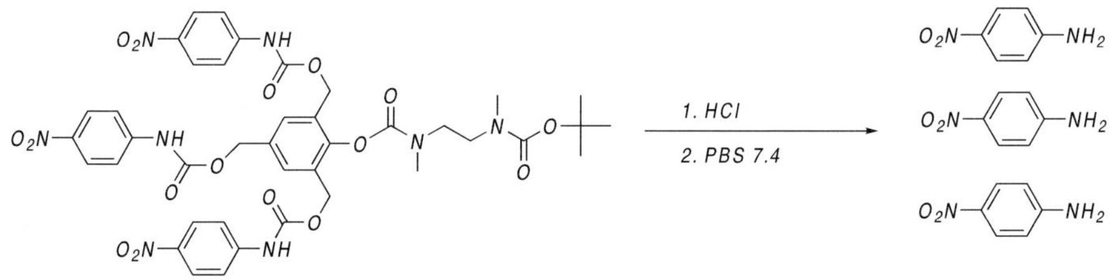

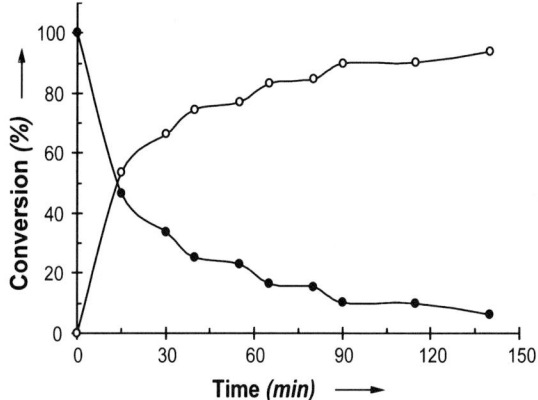

Scheme 23 Boc cleavage triggers the release of *p*-nitroaniline from the trimeric platform

Fig. 8 Release of *p*-nitroaniline from the trimeric platform, (○) *p*-nitroaniline, (●) starting material

saline (PBS) at pH 7.4. HPLC analysis was used to monitor the reaction progress, by following the formation of *p*-nitroaniline. Figure 8 shows that free *p*-nitroaniline was rapidly generated after the cleavage of the Boc protecting group. No release was observed when the Boc group remained attached to the platform.

5.2
Homotrimeric Dendritic Prodrug

The above experiment proved that the triple quinone-methide rearrangement could indeed take place under physiological conditions. With this information in hand, we synthesized a trimeric prodrug system linking three molecules of the anticancer drug camptothecin (pro-tCPT) through a retro-aldol retro-Michael trigger to a substrate for catalytic antibody 38C2. In addition, we synthesized a monomeric CPT prodrug (pro-mCPT) with an identical linker (Fig. 9a). Both prodrugs were activated upon incubation with antibody 38C2 and the CPT release was confirmed by HPLC analysis (data not shown).

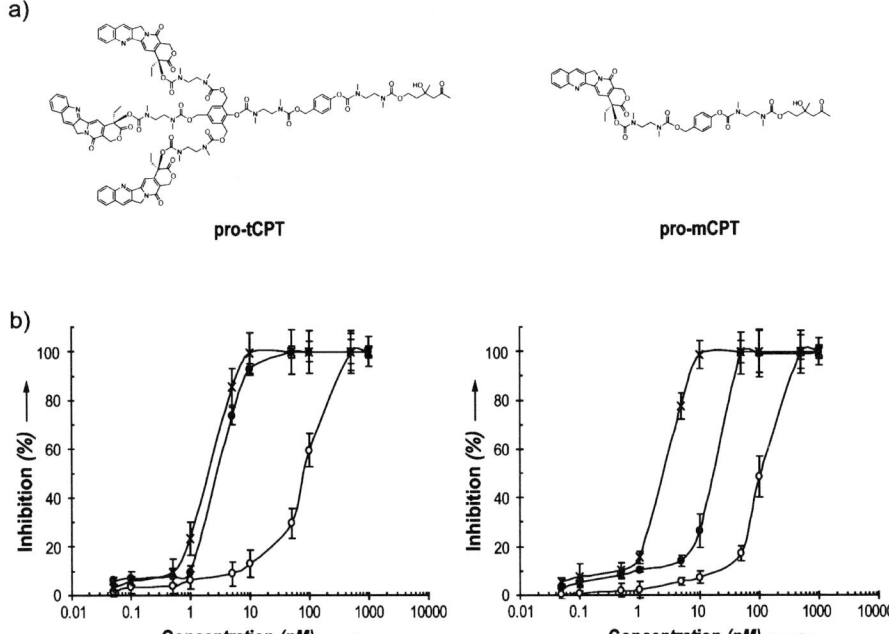

Fig. 9 **a** Molecular structures of a single triggered CPT trimeric prodrug vs. CPT classic monomeric prodrug with identical trigger. **b** Growth inhibition assay of human Molt-3 leukemia cell line, in the presence and absence of catalytic antibody 38C2 *Left*: (o) pro-tCPT, (∗) pro-tCPT + 38C2, (o) CPT. *Right*: (•) pro-mCPT, (•) pro-mCPT + 38C2, (∗) CPT

Next, we examined whether the trimeric prodrug system had an advantage over the monomeric one in a cell growth inhibition assay. We evaluated the ability of the prodrugs to inhibit cell proliferation in the presence of catalytic antibody 38C2 using three different cell lines: the human T-lineage acute lymphoblastic leukemia (ALL) cell line MOLT-3, the human erythroleukemia cell line HEL, and the human acute myeloid leukemia (AML) cell line HL-60. The results are summarized in Table 1 and the data from the Molt-3 cell line are presented in Fig. 9b. The trimeric prodrug is more potent then the monomeric one when incubated with the antibody, as expected, since the total amount of CPT release is tripled in comparison with the release from an equivalent concentration of monomeric prodrug.

5.3
In Vitro Advantage of Dentritic Prodrug vs. the Monomeric One

In the trimeric system, one cleavage by the antibody releases three times the amount of CPT than a cleavage in the monomeric prodrug system. We selected one cell line (Molt-3 leukemia) for further studies with fixed con-

Table 1 IC$_{50}$ [37] Values from cell-growth inhibition assays

Drug/Prodrug	MOLT-3		HL60		HEL	
	IC$_{50}$[a]	IC$_{50}$[b]	IC$_{50}$[a]	IC$_{50}$[b]	IC$_{50}$[a]	IC$_{50}$[b]
CPT	2.2	2.0	9.0	7.5	13	11
pro-mCPT	100	15	150	31	400	100
pro-tCTP	80	2.7	100	7.5	200	19

[a] Cells were incubated in medium with drug/prodrug.
[b] Cells were incubated in medium with drug/prodrug + 1 µM catalytic antibody 38C2

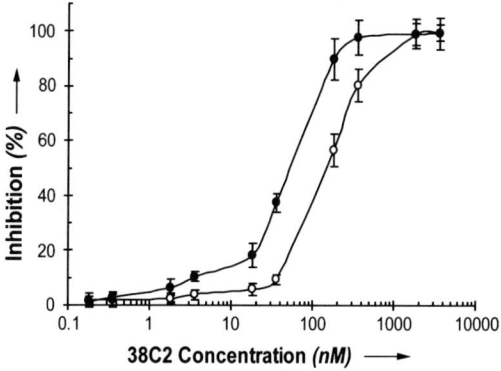

Fig. 10 Growth inhibition assay of the human Molt-3 leukemia cell line, with a fixed concentration of prodrugs and varying concentration of catalytic antibody 38C2. Cells were incubated for 72 h. (o) 36 nM pro-mCPT, (•) 12 nM pro-tCPT

centrations of the prodrug and varying concentrations of antibody 38C2. In order to have equal amounts of CPT, the monomeric prodrug concentration was used at three times the concentration of the trimeric one. The results are shown in Fig. 10. The trimeric prodrug inhibited cell growth up to three times more effectively than the monomeric one in the range of 15–150 nM antibody. In other words, the antibody concentration needed to achieve 50% cell growth inhibition with the pro-tCPT is about three times less than the one used in the pro-mCPT system. It should also be noted that the cytotoxicity of the platform degradation products was previously evaluated in cell growth inhibition. It was found to have negligible or no toxicity at all within the drug concentration range of the cell assay.

5.4
Heterotrimeric Dendritic Prodrug

It is also possible to incorporate three different drug molecules on the same prodrug platform. This would effectively allow triple-drug therapy in a single

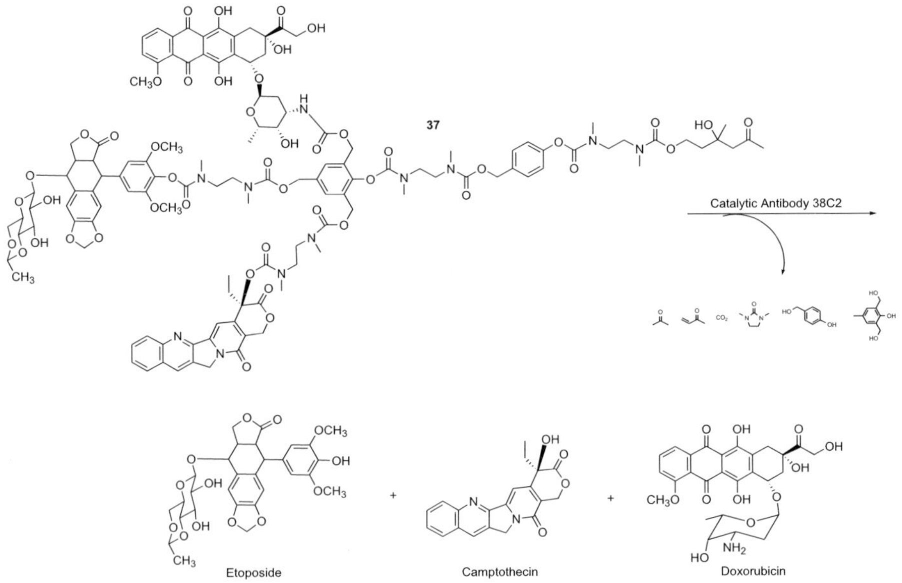

Scheme 24 Single-triggered hetero-trimeric prodrug system with the anti-cancer drugs CPT, doxorubicin, etoposide, and a retro-aldol retro-Michael substrate

molecule. We synthesized a hetero-trimeric system with the anti-cancer drugs CPT, doxorubicin, and etoposide (**37**) using the retro-aldol retro-Michael trigger activated by antibody 38C2 (Scheme 24). Upon single activation cleavage by the catalytic antibody, this prodrug system should almost simultaneously

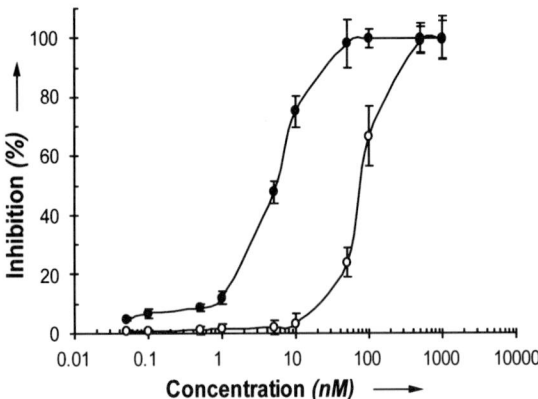

Fig. 11 Growth inhibition assay of the human Molt-3 leukemia cell line, cells were incubated for 72 h. **A** (•) Hetero-trimeric prodrug (pro-37) + 1 μM of catalytic antibody 38C2, **B** (○) pro-37

release three different chemotherapeutic drugs at the same location. HPLC analysis confirmed the release of the drugs in the presence of antibody 38C2.

The hetero-trimeric prodrug system was evaluated in a cell growth inhibition assay (Fig. 11). The prodrug was incubated with the Molt-3 leukemia cells in the presence and in the absence of catalytic antibody 38C2. The cell growth inhibition of the hetero-trimeric prodrug was increased approximately 15-fold upon activation by antibody 38C2. It was previously shown that a hetero-dimeric prodrug system is more effective in inhibition of cell growth than the combination of two monomeric prodrugs [34]. Single-triggered hetero-trimeric prodrugs may offer additional synergy in chemotherapy.

Dendritic prodrugs, activated through a single catalytic reaction by a specific enzyme, could offer significant advantages in inhibiting tumor growth, especially if the targeted or secreted enzyme exists at relatively low levels in the malignant tissue.

6
Linearly Disassembled Dendrons

McGrath also reported on dendrimers that can disassemble linearly in organic solvents by benzyl-ether depolymerization, triggered by an allyl-ether deprotection [40]. The disassembly mechanism occurs through 1,6-elimination to generate a quinone-methide species and an alkoxy leaving group. There is no amplification effect in this example. Zeroth, first, and second generation dendrons were synthesized (Scheme 25) and upon the allyl-ether deprotection, they disassembled linearly to release 4-nitrophenol. The depolymerization was followed conveniently with visible spectroscopy.

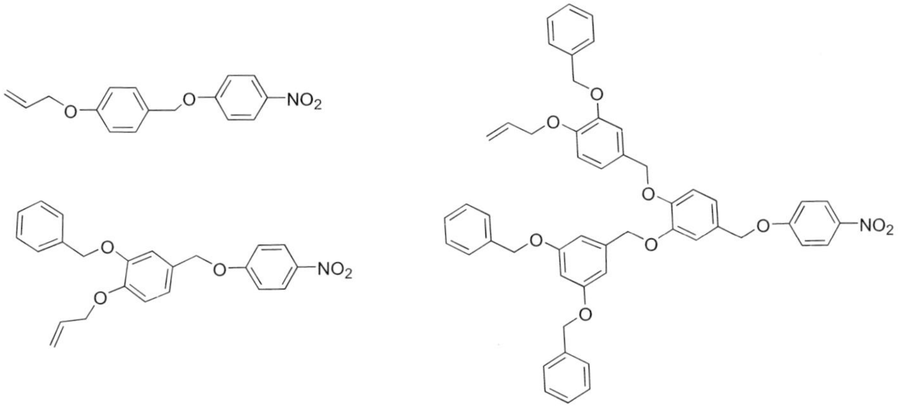

Scheme 25 Chemical structure of zeroth, first, and second generation linearly disassembled dendrons with an allyl trigger and *p*-nitrophenol reporter

7
Multi-Triggered Domino Dendrons

7.1
Mechanism

We extended the above concept to fully biodegradable dendrimers with reasonable solubility in water and disassembled through multi-enzymatic triggering followed by self-immolative chain fragmentation [41]. The dendrimer's main building block is based on diethylenetriamine, which has two primary and one secondary amine functionalities. In a first-generation dendron (Scheme 26), the secondary amine is attached to a reporter group while the two primary amines are linked to enzymatic substrates. The cleavage of either of the substrates by the enzyme, generates a free amine group which initiates an intra-cyclization reaction to release the reporter group.

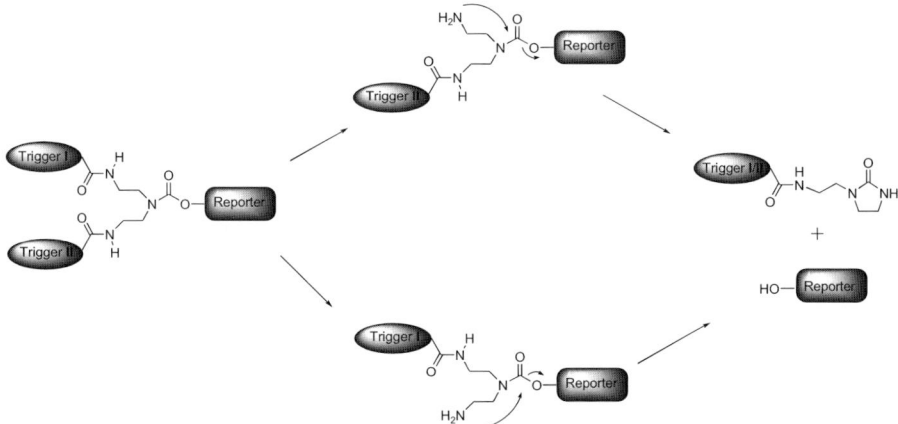

Scheme 26 Description of a G1-dendron disassembly through a double triggering mechanism. Cleavage of either trigger **I** or **II** will initiate the release of the reporter group

7.2
Dendron Structure

In order to evaluate our dendrimer biodegradation pathway, we synthesized zeroth, first, and second-generation dendrons (Scheme 27) with phenylacetamide as a triggering substrate for penicillin-G-amidase (PGA) and 4-nitrophenol as a reporter group. 4-Hydroxybenzyl alcohol was employed as a self-immolative linker to connect between two amine groups through carbamate linkages.

Similar to a first-generation dendron, a second-generation dendron (compound **40**) can disassemble to its building blocks through the described enzy-

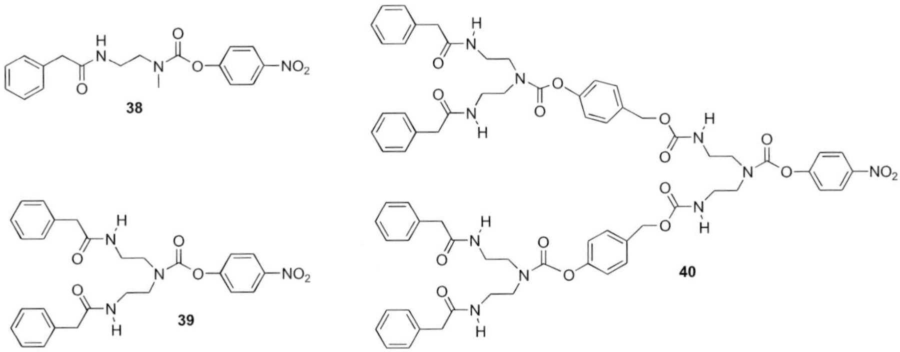

Scheme 27 Chemical structure of zeroth, first, and second-generation dendrons

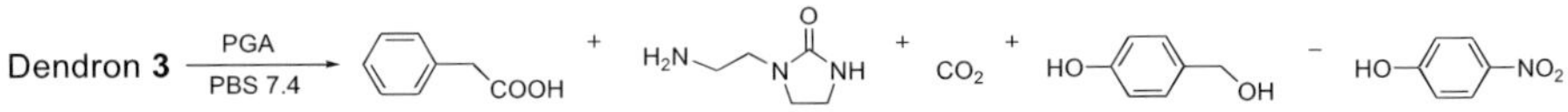

Scheme 28 PGA-catalyzed fragmentation of a G3-dendron to its building blocks

matic self-immolative fragmentation. The phenol, which is released after the first intra-cyclization, undergoes 1,6-quinone-methide rearrangement to release carbamic acid from the benzylic carbon. The quinone-methide species is rapidly trapped by a water molecule to yield 4-hydroxybenzyl alcohol. The generated carbamic acid, undergoes spontaneous decarboxylation to form a free amine group, which is self-cyclized to release the reporter group. Importantly, only one enzymatic cleavage, out of a possible four, is sufficient to initiate the domino breakdown that will release the reporter group at the focal point of the dendrimer. The complete degradation of the dendron to its building blocks is depicted in Scheme 28.

7.3
Enzymatic Activation of a Multi-Triggered Dendron

Dendrons **38–40** were incubated with PGA in PBS pH 7.4 at 37 °C. Their biodegradation was conveniently monitored by following the formation of 4-nitrophenol with visible spectroscopy at a wavelength of 405 nm. The kinetic release of 4-nitrophenol from the dendrons is shown in Fig. 12. Upon addition of PGA to dendrons **38–40**, free 4-nitrophenol was gradually formed, indicating that PGA cleaves its phenylacetamide substrate and the degradation indeed occurs as was predicted. As expected first-generation dendron released the 4-nitrophenol faster than zeroth-generation dendron while second-generation dendron released it more slowly. The background control reactions showed no release at all.

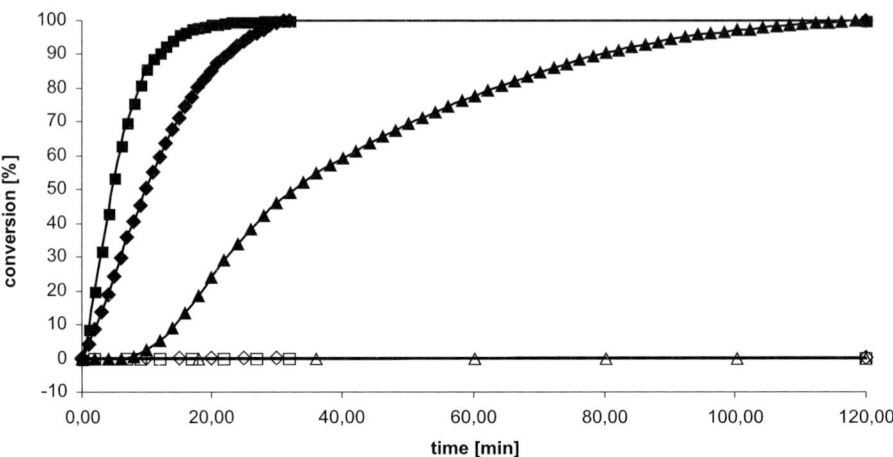

Fig. 12 UV Absorbance at 405 nm as a function of time in the biodegradation of the self-immolative dendrons. ♦ dendron **38** + PGA. ■ dendron **39** + PGA. ▲ dendron **40** + PGA. □ dendron **38** in PBS pH 7.4. ◇ dendron **39** in PBS 7.4. △ dendron **40** in PBS pH 7.4 (substrate concentration is 200 μM and 10 μM for PGA)

The kinetic constants $K_{(obs)}$ for the three reactions were calculated by linear correlation with the measured plots (Table 2). The phenomenon of dendron **39** releasing its reporter group faster than dendron **38** occurs since the enzymatic substrate concentration in dendron **39** is twice that of dendron **38**. The followed self-cyclization step is relatively fast and therefore the rate-limiting step is cleavage of the enzymatic substrate. In dendron **40**, additional self-immolative reactions occurred in order to complete the release of the reporter group (another intra-cyclization and 1,6-quinone-methide elimination). The overall rate of these reactions is slower than the rate of the enzymatic substrate cleavage and therefore the $K_{(obs)}$ for dendron **40** is relatively smaller.

In this study, we designed and synthesized new dendritic molecules with a multi-enzymatic triggering mechanism that initiates their biodegradation through a self-immolative chain fragmentation to release a reporter group from the focal point. For the first time, the potential of diethylenetriamine was introduced as a double trigger linker, which can be used as a building block for constructing self-immolative dendrimers. The dendrons were

Table 2 $K_{(obs)}$ values for the reporter release reactions for dendrons **38–40**

	Dendron **38**	Dendron **39**	Dendron **40**
$K_{(obs)}$ [min^{-1}]	5.11	9.89	2.43

found to have fairly good (zeroth and first-generation) to moderate (second generation) water solubility and high stability to background hydrolysis under physiological conditions. Their degradation readily occurs in aqueous medium and can easily be monitored by generation of free reporter molecules. Incorporation of different substrates in the dendron's periphery, should allow the use of diverged triggering enzymes [42]. This concept may be particularly important in the field of prodrug mono-therapy, if a drug molecule is incorporated instead of the reporter unit, especially, in circumstances with more than one tumor-associated or targeted enzyme with different catalytic activity.

8
Outlook

A highlight that appeared recently in "Nature" begun with the sentence, "If a good idea for scientific innovation emerges, you can be sure that several teams of researchers will be quickly on the case." [43]. Surprisingly, our group and two others reported almost simultaneously, the domino dendrimers concept, independently. The three groups designed and synthesized a new class of dendritic molecules that disassemble through a domino-like mechanism. These structurally unique dendrimers can release all their tail units through a self-immolative chain fragmentation, which is initiated by a single cleavage at the dendrimer's core. One possible intriguing application for these new molecules, could be accomplished in the form of dendritic prodrugs, which are activated through a single catalytic reaction by a specific enzyme, over-expressed in a tumor tissue. Dendritic prodrugs could offer significant advantages in the inhibition of tumor growth, especially if the targeted or secreted enzyme exists at relatively low levels in the malignant tissue. De Groot addressed this concept with the synthesis of dendritic taxol prodrug. However, it was not activated under physiological conditions. We were able to achieve the bioactivation of dimeric and trimeric-dendritic prodrugs. It was also shown that dendritic prodrug exhibits a clear advantage over the classic monomeric one, in a cancerous cell-growth inhibition assay. The next step required in order to advance this application, should be an in-vivo evaluation of a dendritic prodrug, which is selectively activated by a tumoral enzyme. In order to achieve this goal, one should try to increase the water solubility of the dendritic molecules and reach an appropriate concentration for in vivo studies.

Domino dendrimers may also be applied as a general platform for sensor molecules used to detect enzymatic activity. The reporter units could be any molecule with a hidden signal, which is revealed upon the unit release. The dendritic platform will act as a molecular enhancer, which amplifies a single enzymatic activation into multi-detectable signals.

References

1. Tomalia DA, Frechet JMJ (2002) Journal of Polymer Science: Polymer Chemistry 40:2719–2728
2. Grinstaff MW (2002) Biodendrimers: new polymeric biomaterials for tissue engineering. Chemistry 8:2839–2846
3. Stiriba S-E, Frey H, Haag R (2002) Angew Chem Int Ed 41:1329–1334
4. Patri AK, Majoros IJ, Baker JR (2002) Dendritic polymer macromolecular carriers for drug delivery. Curr Opin Chem Biol 6:466–471
5. Kim Y, Zimmerman SC (1998) Applications of dendrimers in bio-organic chemistry. Curr Opin Chem Biol 2:733–742
6. Tomalia DA (2004) Birth of a new macromolecular architecture: Dendrimers as quantized building blocks for nanoscale synthetic organic chemistry. Aldrichimica Acta 37:39–57
7. Aulenta F, Hayes W, Rannard S (2003) Dendrimers: a new class of nanoscopic containers and delivery devices. Eur Poly J 39:1741–1771
8. Boas U, Heegaard PMH (2004) Dendrimers in drug research. Chem Soc Rev 33:43–63
9. Bosman AW, Janssen HM, Meijer EW (1999) About dendrimers: structure, physical properties, applications. Chem Revs (Washington, DC) 99:1665–1688
10. Kobayashi H, Brechbiel MW (2003) Dendrimer-based macromolecular MRI contrast agents: characteristics and application. Mol Imag 2:1–10
11. D'Emanuele A, Jevprasesphant R, Penny J, Attwood D (2004) The use of a dendrimer-propranolol prodrug to bypass efflux transporters and enhance oral bioavailability. J Contr Rel 95:447–453
12. Radu DR, Lai C-Y, Jeftinija K, Rowe EW, Jeftinija S, Lin VSY (2004) A polyamidoamine dendrimer-capped mesoporous silica nanosphere-based gene transfection reagent. J Am Chem Soc 126:13216–13217
13. Kim T-I, Seo HJ, Choi JS, Jang H-S, Baek J, Kim K, Park J-S (2004) PAMAM-PEG-PAMAM: novel triblock copolymer as a biocompatible and efficient gene delivery carrier. Biomacromolecules 5:2487–2492
14. Kukowska-Latallo JF, Bielinska AU, Johnson J, Spindler R, Tomalia DA, Baker JR Jr (1996) Efficient transfer of genetic material into mammalian cells using Starburst polyamidoamine dendrimers. Proceedings of the National Academy of Sciences of the United States of America 93:4897–4902
15. Patri AK, Myc A, Beals J, Thomas TP, Bander NH, Baker JR Jr (2004) Synthesis and in vitro testing of J591 antibody-dendrimer conjugates for targeted prostate cancer therapy. Biocon Chem 15:1174–1181
16. Thomas TP, Patri AK, Myc A, Myaing MT, Ye JY, Norris TB, Baker JR Jr (2004) In vitro targeting of synthesized antibody-conjugated dendrimer nanoparticles. Biomacromolecules 5:2269–2274
17. Fuchs S, Kapp T, Otto H, Schoeneberg T, Franke P, Gust R, Schlueter AD (2004) A surface-modified dendrimer set for potential application as drug delivery vehicles: synthesis, in vitro toxicity, and intracellular localization. Chemistry—A European Journal 10:1167–1192
18. Ooya T, Lee J, Park K (2004) Hydrotropic dendrimers of generations 4 and 5: synthesis, characterization, and hydrotropic solubilization of paclitaxel. Bioconjugate Chemistry 15:1221–1229
19. Kawano T, Okuda T, Aoyagi H, Niidome T (2004) Long circulation of intravenously administered plasmid DNA delivered with dendritic poly(-lysine) in the blood flow. Journal of Controlled Release 99:329–337

20. Paleos CM, Tsiourvas D, Sideratou Z, Tziveleka L (2004) Acid- and salt-triggered multifunctional poly(propylene imine) dendrimer as a prospective drug delivery system. Biomacromolecules 5:524–529
21. Mammen M, Chio S-K, Whitesides GM (1998) Polyvalent interactions in biological systems: implications for design and use of multivalent ligands and inhibitors. Angew Chem Int Ed 37:2755–2794
22. Shaunak S, Thomas S, Gianasi E, Godwin A, Jones E, Teo I, Mireskandari K, Luthert P, Duncan R, Patterson S, Khaw P, Brocchini S (2004) Polyvalent dendrimer glucosamine conjugates prevent scar tissue formation. Nat Biotech 22:977–984
23. Benito JM, Gomez-Garcia M, Ortiz Mellet C, Baussanne I, Defaye J, Fernandez JMG (2004) Optimizing saccharide-directed molecular delivery to biological receptors: design, synthesis, and biological evaluation of glycodendrimer-cyclodextrin conjugates. J Am Chem Soc 126:10355–10363
24. Ihre HR, Padilla De Jesus OL, Szoka FC Jr, Frechet JMJ (2002) Polyester dendritic systems for drug delivery applications: design, synthesis, and characterization. Biocon Chem 13:443–452
25. Amir RJ, Pessah N, Shamis M, Shabat D (2003) Self-immolative dendrimers. Angew Chem Int Ed 42:4494–9
26. Shabat D, List B, Amir RJ, Shamis M, Pessah N (2004) In: PCT Int Appl p 153 (Ramot at Tel Aviv University Ltd., Israel; The Scripps Research Institute) Wo
27. de Groot FM, Albrecht C, Koekkoek R, Beusker PH, Scheeren HW (2003) "Cascade-release dendrimers" liberate all end groups upon a single triggering event in the dendritic core. Angew Chem Int Ed 42:4490–4494
28. Szalai ML, Kevwitch RM, McGrath DV (2003) Geometric disassembly of dendrimers: dendritic amplification. J Am Chem Soc 125:15688–15689
29. Szalai ML, McGrath DV (2004) Phototriggering of geometric dendrimer disassembly: an improved synthesis of 2,4-bis(hydroxymethyl)phenol based dendrimers. Tetrahedron 60:7261–7266
30. Firestone R (1998) In: PCT Int Appl p 63 (Bristol-Myers Squibb Co, USA). Wo
31. Flomenbom O, Amir RJ, Shabat D, Klafter J (2005) Some new aspects of dendrimer applications. J of Luminescence 111:315–325
32. de Groot FM, Damen EW, Scheeren HW (2001) Anticancer prodrugs for application in monotherapy: targeting hypoxia, tumor-associated enzymes, and receptors. Curr Med Chem 8:1093–1122
33. Bagshawe KD, Springer CJ, Searle F, Antoniw P, Sharma SK, Melton RG, Sherwood RF (1988) A cytotoxic agent can be generated selectively at cancer sites. Br J Cancer 58:700–703
34. Shamis M, Lode HN, Shabat D (2004) Bioactivation of self-immolative dendritic prodrugs by catalytic antibody 38C2. J Am Chem Soc 126:1726–1731
35. Wagner J, Lerner RA, Barbas CF III (1995) Efficient aldolase catalytic antibodies that use the enamine mechanism of natural enzymes. Science (Washington, DC) 270:1797–1800
36. Shabat D, Rader C, List B, Lerner RA, Barbas CF III (1999) Multiple event activation of a generic prodrug trigger by antibody catalysis. Proc Natl Acad Sci USA 96:6925–6930
37. Rader C, Turner JM, Heine A, Shabat D, Sinha SC, Wilson IA, Lerner RA, Barbas CF (2003) A humanized aldolase antibody for selective chemotherapy and adaptor immunotherapy. J Mol Biol 332:889–899
38. Shabat D, Lode H, Pertl U, Reisfeld RA, Rader C, Lerner RA, Barbas CF III (2001) In vivo activity in a catalyic antibody-prodrug system: Antibody catalyzed etoposide prodrug activation for selective chemotherapy. Proc Natl Acad Sci USA 98:7528–7533

39. Haba K, Popkov M, Shamis M, Lerner RA, Barbas CF III, Shabat D (2005) Single-triggered trimeric prodrugs. Angew Chem Int Ed 44:716–720
40. Li S, Szalai ML, Kevwitch RM, McGrath DV (2003) Dendrimer disassembly by benzyl ether depolymerization. J Am Chem Soc 125:10516–10517
41. Amir RJ, Shabat D (2004) Self-immolative dendrimer biodegradability by multi-enzymatic triggering. Chem Commun (Camb) 14:1614–1615
42. Gopin A, Pessah N, Shamis M, Rader C, Shabat D (2003) A chemical adaptor system designed to link a tumor-targeting device with a prodrug and an enzymatic trigger. Angew Chem Int Ed 42:327–332
43. Meijer EW, van Genderen MHP (2003) Chemistry: Dendrimers set to self-destruct. Nature (London, United Kingdom) 426:128–129

PEGylation of Proteins as Tailored Chemistry for Optimized Bioconjugates

Gianfranco Pasut · Francesco M. Veronese (✉)

Department of Pharmaceutical Sciences, University of Padua, via F. Marzolo 5, 35131 Padova, Italy
gianfranco.pasut@unipd.it, francesco.veronese@unipd.it

1	Introduction	96
2	Features of PEG as a Bioconjugation Polymer	98
2.1	PEGs for Protein Modification	100
2.2	Forthcoming PEGs	105
3	Challenges in Protein PEGylation	105
4	PEGylation Chemistry	108
4.1	PEGylation on Protein Amino Groups	108
4.1.1	Interferons	109
4.1.2	Granulocyte Colony Stimulating Factor (G-CSF)	112
4.1.3	Megakaryocyte Growth and Development Factors (MGDFs)	114
4.1.4	Growth Hormone (GH), Growth Hormone-releasing Hormone (GRF), and Growth Hormone Antagonist	115
4.1.5	Antibodies and Antibody Fragments	118
4.1.6	Others Proteins	120
4.2	PEGylation on the Protein Thiol Groups	122
4.3	Protein PEGylation Catalyzed by Enzymes	125
4.4	Protein PEGylation of Carboxylic Groups	127
4.5	Beyond Protein PEGylation	128
5	Conclusion	129
	References	131

Abstract The high potential of peptides and proteins as therapeutic agents has not been fully exploited because of their common shortcomings: the only exist for a short lifetime in the body, they degrade easily in vivo and in vitro, and they cause immunological reactions. Among several proposed solutions, PEGylation, the covalent modification using polyethylene-glycol (PEG), has achieved interesting results, leading to a novel series of products that have already reached the market, while others will be available soon. In the past few years this technology, first developed for peptides and proteins, has been applied to non-peptide drugs, opening a new area of investigation that is receiving increasing interest. In this case, PEGylation allows the therapeutic application of molecules with inadequate water solubility, high toxicity, or a poor pharmacokinetic profile. This chapter describes recent achievements in PEGylation of proteins and peptides, with a special emphasis on the chemistry of conjugation, and it reports many examples from literature and from the authors' own experimental results.

Keywords PEG · PEGylation · PEG-protein · PEG-drugs · Polymer therapeutics

1
Introduction

In the past decade drug delivery systems (DDSs) have become an interesting and fast-growing research field to which many pharmaceutical companies are looking in order to improve their products, those being either small or large molecules like peptides, proteins, non-protein drugs, and polynucleotides. Using different strategies, the aim of a DDS is to improve the pharmacokinetic and pharmacodynamic profiles of a therapeutic agent.

The most investigated strategies are drug-polymer conjugation, drug incorporation into micro/nanoparticles, polymeric matrices, liposomes, and micelles. Some of the advantages that one may achieve include controlled drug release, reduced body clearance, increased stability, lower toxicity, and enhanced specificity and efficacy. A suitable DDS is sometimes necessary as in the case of a drug with low solubility or a narrow therapeutic index.

Polymer bioconjugation plays an important role in the field of DDSs, mainly for drugs having high molecular weight and a peptide structure, but also in the case of a special class of therapeutic agent, such as anti-cancer agents, which usually have low molecular weight with a non-peptide structure. The success of this technology is reflected not only by the several drug-polymer conjugates already on the market or under advanced clinical investigation, but also by the growing number of publications and patents appearing each year.

A lot of polymers from both natural and synthetic sources have been considered for modification. Polysaccharides are an example of the former, while polyacrylate copolymers and PEG are examples of the latter.

While all drugs already on the market or under development may be candidates for alternative delivery methods, it is noteworthy that the peptide drugs are the most likely candidates, since they are worth more than $10 billion within the world pharmaceutical market and represent a rapidly growing segment. As drug delivery is closely tied with pharmaceutical manufacture, it is anticipated that its market will be worth an estimated $120 billion by 2007, and bioconjugation appears to be one of the most promising approaches to reach this goal.

The research in the field of protein modification with polymers started in the 1960s and 1970s with dextran as the polymer. However, a real boost in this field occurred with the introduction of PEG thanks to the pioneering studies conducted in the late 1970s by Professor Frank Davis at Rutgers University [1]. Since then, many studies focusing on the development of polymer conjugation chemistry, analytical investigation, and purification techniques have been conducted. A number of drug candidates of different structure were PEGylated, including proteins, peptides, low-molecular-weight drugs, and polynucleotides, as was reported in many publications and patents [2].

PEG is the polymer of choice for protein modification because it possesses several favorable properties such as the lack of immunogenicity, antigenicity, and toxicity, and a high solubility in water and in many organic solvents—it is also approved by the FDA for human use. Common reasons for the PEGylation of a drug are to reduce its excretion by the kidneys, to avoid or reduce its degradation by proteolytic enzymes and/or hydrolytic media, to enhance its water solubility (for highly hydrophobic molecules), to reduce its reticuloendothelial (RES) clearance, and to reduce its immunogenicity and antigenicity (mainly for peptides and proteins) [3–8]. Furthermore, for small drugs, polymer conjugation may yield improved and more convenient biodistribution, selected cellular uptake [9–11] or, through tailor-made chemistry, a triggered drug release or targeting into specific organs or cells [12]. Advantages of PEGylation are summarized in Table 1 and reported in several reviews [13–15]. These goals are achieved by a combination of an increase in the molecular weight of conjugates, the coverage or blockage of selected protein sites (epitopes or sequences degraded by enzymes), and a high polymer solubility in water.

The history of protein PEGylation may be divided into two generations:

- The first generation of conjugates refers to PEGs with low molecular weight (≤ 12 kDa) and to monomethoxy PEG (mPEG) batches having a relevant percentage of diol chains (originating from polymer synthesis)—an impurity that is a potential cross-linking agent. The chemistry employed for mPEG-protein conjugation often presented side reaction products or led to weak and reversible linkages. Despite these initial difficulties, important products were created, and some reached the market, such as PEG-adenosine deaminase (Adagen®) [16] for the treatment of severe combined immunodeficiency disease (SCID), and PEG-asparaginase (Oncaspar®) [17] for the treatment of leukemia.

Table 1 General advantages of bioconjugation in therapeutic applications

Stabilization of labile drugs from chemical degradation
Protection from proteolytic degradation
Reduction of immunogenicity
Decreased antibody recognition
Increased body residence time
Modification of biodistribution
Drug penetration by endocytosis
New strategies for drug targeting
Increased water solubility
Reduced toxicity

- The second generation is represented by an improvement in PEG purity, a reduction in both polydispersivity and diol content (also at the industrial scale), a greater selectivity for protein modification, and a large range of available activated PEGs. Among the new PEGs, the branched variety has a wider range of applications due to its enhanced size compared to the common linear PEGs. Heterobifunctional PEGs were also prepared in order to link a second molecule with a targeting role; spacers, between the polymer and the drug, were studied as either reporter groups or to allow the release of a bound drug under specific triggering conditions. Several products of this second generation have reached the market, including linear PEG-interferon α2b (PEG-Intron®) [18], branched PEG-interferon α2a (Pegasys®) [19, 20], PEG-growth hormone receptor antagonist (Pegvisomant, Somavert®) [21], PEG-G-CSF (pegfilgrastim, Neulasta®) [22], and branched PEG-anti-VEGF aptamer (Pegaptanib sodium injection, Macugen™) [23]; many others are presently under clinical trials and hopefully will be available in the near future.

2
Features of PEG as a Bioconjugation Polymer

Raw poly(ethylene glycol) is synthesized by ring-opening polymerization of ethylene oxide. The reaction is initiated by methanol or water, forming polymers with one or two end-chain hydroxyl groups, respectively (mPEG–OH or HO – PEG – OH). Starting from these simple forms, a large series of PEG derivatives were developed to address chemical groups with different reactivity in the drug molecule. PEGs with various shapes and functionalization are now commonly available: branched, multifunctional, and several heterobifunctional PEGs (Fig. 1). Monofunctional polymers (mPEG–OH), linear or branched, are particularly indicated for protein modification, while those with multiple groups are useful for enhancing the loading of low-molecular-weight drugs, especially for those with reduced biological activity that otherwise would need the administration of a large amount of polymer.

The development of a well-controlled polymerization procedure and an adequate purification process is now leading to low-polydisperse polymer batches, M_w/M_n spanning from 1.01 for PEG below 5 kDa in molecular weight up to 1.1 for PEG as high as 50 kDa.

PEG has unique solvation properties that are due to the coordination of 2–3 water molecules per ethylene oxide unit [2] that, together with the great flexibility of the polymer backbone, are responsible for the protein-rejecting properties of PEG and the biocompability, which form the basis of the antimmunogenicity and antigenicity conveyed to the conjugates [24]. Furthermore, these characteristics give PEG molecules an apparent molecular weight 5–10

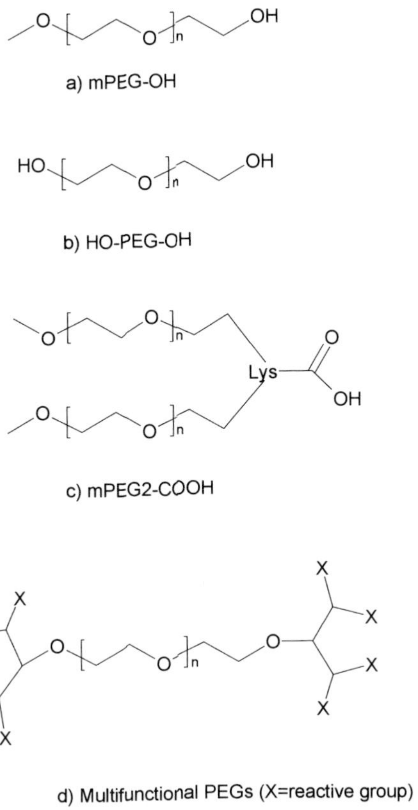

Fig. 1 Different PEG structures: **A** linear monomethoxy PEG, **B** linear diol PEG, **C** branched PEG, and **D** multifunctional PEGs

times higher than that of a globular protein of comparable mass, as verified by gel permeation chromatography [25]. In this large hydrodynamic volume, PEG covers, by steric hindrance, an extended surface of the conjugated protein, preventing degradation by mammalian cells and enzymes [26].

In vivo, PEG undergoes limited chemical degradation, and its clearance depends upon its molecular weight: below 20 kDa it is easily secreted into urine, while at higher molecular weight it is eliminated more slowly, and clearance through the liver becomes predominant. The threshold for urine elimination of proteins is approximately 40–60 kDa (a hydrodynamic radius of approximately 45 Å [27]), which represents the albumin excretion limit. Over this limit the polymer remains in circulation, and it is mainly accumulated in liver, while alcohol dehydrogenase can degrade low-molecular-weight PEGs; chain cleavage can be catalyzed by P450 microsomial enzymes [28]. Molecular weight reduction may also take place throughout chain cleavage, albeit more slowly, as it happens to the polymer after a long period storage, or in branched PEGs where the hydrolysis and loss of one polymer chain is catalyzed by anchimeric assistance [29]. Finally the success of many years of PEG use as an excipient in foods, cosmetics, and pharmaceuticals, without toxic effects, is clear proof of its safety [26].

2.1
PEGs for Protein Modification

The first generation of PEG, as a bioconjugation polymer, was studied for amino group modification in proteins. These groups are well represented in proteins, and the chemistry for their modification has been well developed through several studies. A number of activated PEG derivatives have therefore been developed, and, among these, the best known are: a) PEG succinimidyl succinate (SS-PEG), b) PEG succinimidyl carbonate (SC-PEG), c) PEG p-nitrophenyl carbonate (pNPC-PEG), d) PEG benzotriazolyl carbonate (BTC-PEG), e) PEG trichlorophenyl carbonate (TCP-PEG), f) PEG carbonylimidazole (CDI-PEG), g) PEG tresylate, and h) PEG dichlorotriazine (Fig. 2).

The difference among the above reported PEGs lies in the resulting chemical link between the polymer and the drug or in the rate of coupling. The derivatives with slower reactivity, such as carbonate PEGs like pNPC-PEG, CDI-PEG, and TCP-PEG, allow a certain degree of selective conjugation within the amino groups present in a protein according to their nucleophilicity or accessibility [3]. A great difference in reactivity is usually observed between the α and the ε amino group in proteins due to their pKa; in fact, the ε-amino residue of lysine has a pKa of 9.3–9.5, and it is more reactive at high pH than the α-amino group (pKa of 7.6–8.0). Hence the less reactive PEGs will preferentially react with the ε-amino residue of lysine at high pH. By contrast, low pH values (5.5–6.5) leave only the α-amino groups partially unprotonated and still reactive, allowing a selective protein modification [22].

It is noteworthy that the conjugation performed using PEG dichlorotriazine, PEG tresylate, and PEG aldehyde (the last of these after sodium borohydride reduction) maintains the same total charge on the native protein surface, since these derivatives react through an alkylation to form a secondary amine. In contrast, PEGylation conducted with acylating PEGs (i.e., SS-PEG, SC-PEG, pNPC-PEG, CDI-PEG and TCP-PEG) yields to weak acidic amide or carbamate linkages.

The amino groups are the most reactive entities for the above reported PEG derivatives, but PEGs such as SC-PEG, BTC-PEG, and PEG-dichlorotriazine can slowly react both with hydroxyl groups (Ser, Thr, Tyr) and the histidine secondary amino group, giving linkages that are generally hydrolytically unstable. The pH conditions may enhance the percentage of these unconventional PEGylation reactions; for example, α-interferon was conjugated to SC-PEG or BTC-PEG at the His34 side chain under slightly acidic conditions [30] (the pKa value of histidine lies between those of α and

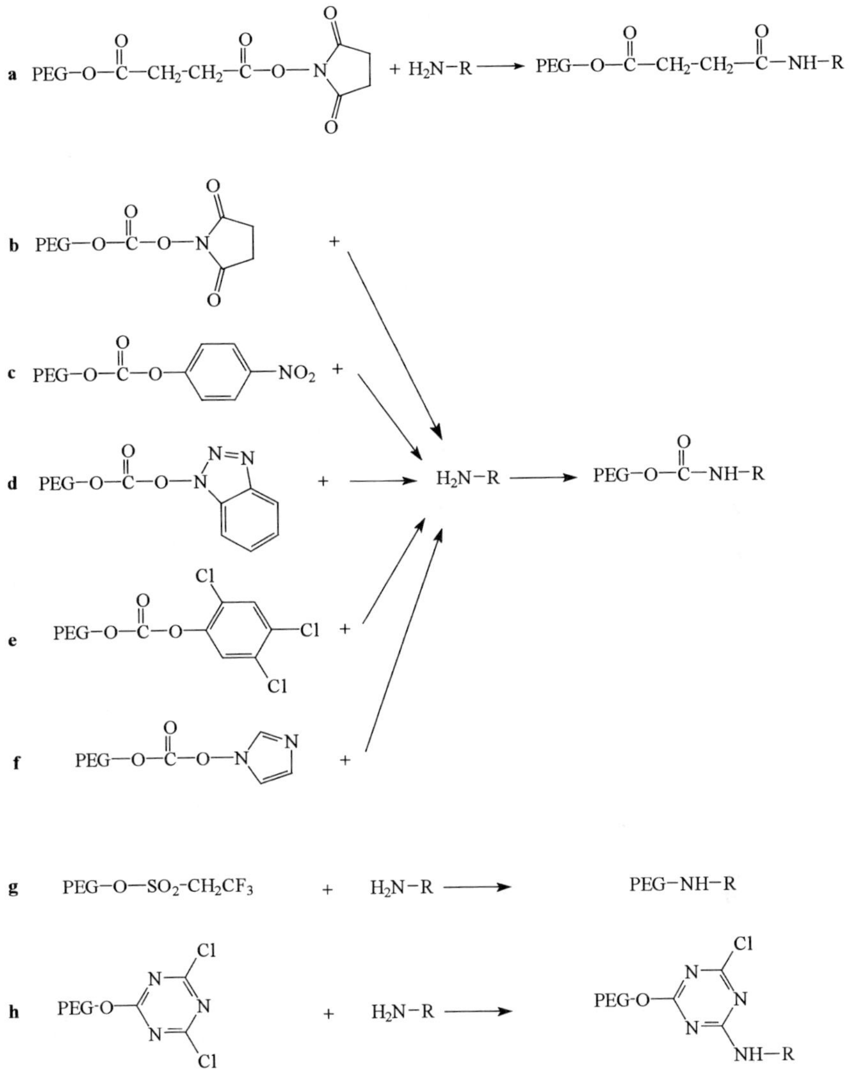

Fig. 2 Examples of activated PEG molecules reactive towards amino groups: **a** PEG succinimidyl succinate, **B** PEG succinimidyl carbonate, **C** PEG *p*-nitrophenyl carbonate, **D** PEG benzotriazol carbonate, **E** PEG trichlorophenyl carbonate, **F** PEG carbonylimidazole, **G** PEG tresylate, and **h** PEG dichlorotriazine

ε amino), and PEG was linked to the serine and tyrosine hydroxyl groups of the decapeptide antide and of epidermal growth factor (EGF) [31, 32], respectively.

As reported above, the first generation of PEGs faced the limitation that there existed quite high percentages of PEG diol contaminant in the methoxy

PEG batches, resulting in a non-negligible amount of dimerized polymer molecules with double molecular weight. This inconvenience was finally solved by the isolation of the mono-carboxylic acid intermediate of mPEG from the bicarboxylic one, coming from the diol, by ionic exchange chromatography. The reduction in PEG diol content, together with the improvement of polymer synthesis and PEGylation chemistry and analysis, yielded to the second generation of polymers, which is still expanding the potential of the technique.

Among the new PEGs that have already marked the evolution of the second generation and that will be important in the near future, one may report:

- PEG-propionaldehyde, also in the form of the more stable acetal: the reaction with an amino group leads to a Shiff base that is reduced by NaCNBH$_3$, giving a derivative that maintains the same net charge of the parent drug;
- PEG-succinimidyl derivatives: highly reactive with respect to amine groups. The reaction rate of these derivatives may significantly change depending upon the extension and the composition of the alkyl chain between a PEG and a succinimidyl moiety [33];
- "Y"-shaped branched PEG [34] (see Fig. 1c): as a result of its increased surface shielding, this PEG reagent is more effective in protecting the conjugated protein from degradative enzymes and antibodies (Fig. 3). Moreover, enzymes modified with this PEG retain more activity with respect to the same enzyme modified by linear PEGs. This effect is probably due to the hindrance of the branching polymer that prevents the entrance of PEG inside the enzyme's active site cleft (Fig. 4);

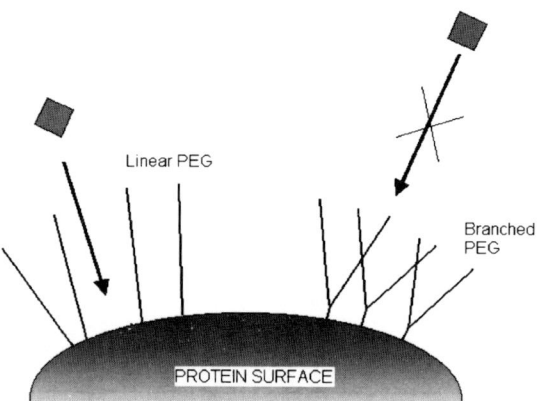

Fig. 3 Structure of linear and branched PEGs on a protein surface. The umbrella-like structure of branched PEG explains the higher capacity in rejecting approaching molecules or cells compared to linear PEG

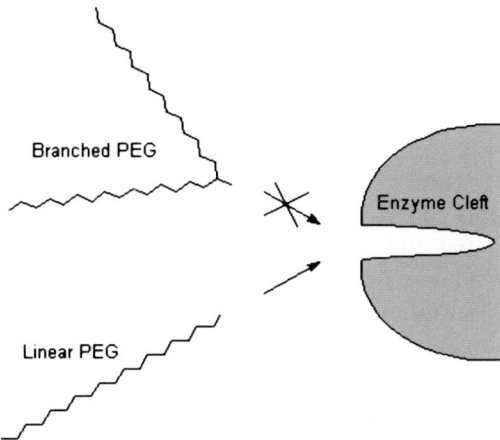

Fig. 4 Effect of PEG hindrance on enzyme active site access. The high steric hindrance of branched PEG may explain the lower deactivation of enzymes as compared to linear PEGs of the same size

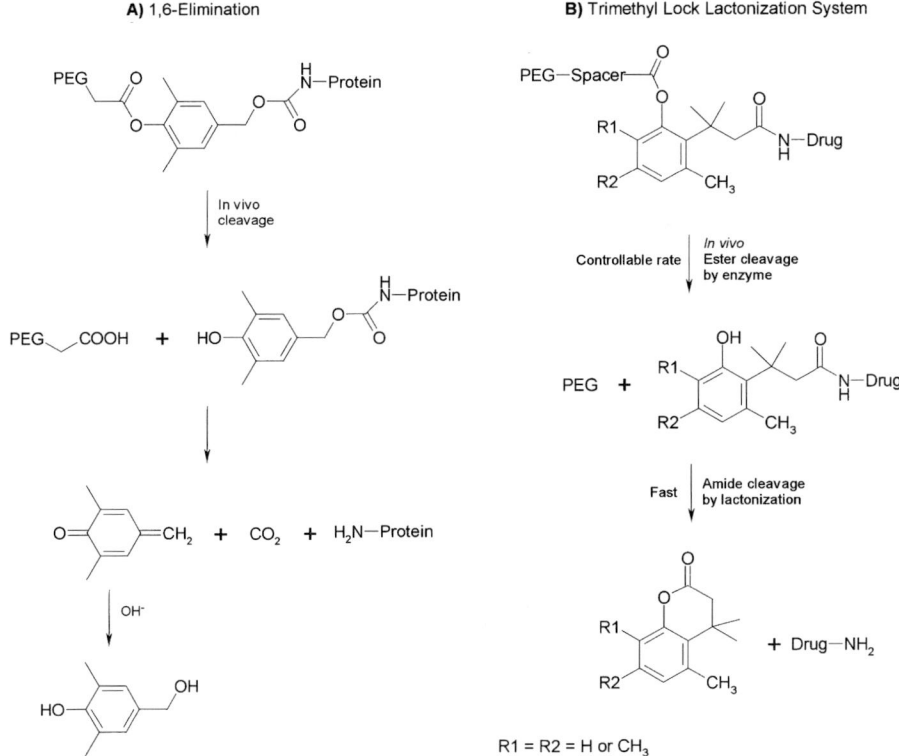

Scheme 1 Controlled release of active molecules from PEGs based on **A** a 1,6-elimination system and **B** a trimethyl lock lactonization system

- PEGs reactive toward thiol groups—PEG-maleimide (MAL-PEG), PEG-vinylsulfone (VS-PEG), and PEG-orthopyridyl-disulfide (OPSS-PEG): even if the thiol addition rate of MAL- or VS-PEGs is very rapid, some addition to amino groups (present in all proteins) may also take place, especially at high pH. On the other hand, the reaction with OPSS-PEG is very specific for thiol groups, but the conjugates may be reversed in the presence of thiols as a reducing agent.
- Heterobifunctional PEGs [35, 36]: these derivatives present two different functional groups, one for each extreme, allowing the linking of two different molecules to the same PEG chain. It is therefore possible to obtain conjugates that carry both a drug and a targeting molecule. Among the proposed and commercially available heterobifunctional PEGs, the ones mainly used are $H_2N - PEG - COOH$, $HO - PEG - COOH$, and $H_2N - PEG - OH$;
- PEG with linkers designed for a controlled release of the conjugated drug: one of the most exploited linkers is a peptide sequence designed to be recognized and cleaved by lysosomal enzymes once the conjugates reach the intracellular compartment. Examples of such peptide linkers may be $H - Gly - Phe - Leu - Gly - OH$ or $H - Gly - Leu - Phe - Gly - OH$ [37, 38]. Alternatively, a linker may respond to pH changes. Moreover, the linker and the polymer together can form a double prodrug system, where the drug release is obtained after polymer hydrolysis (first prodrug), which

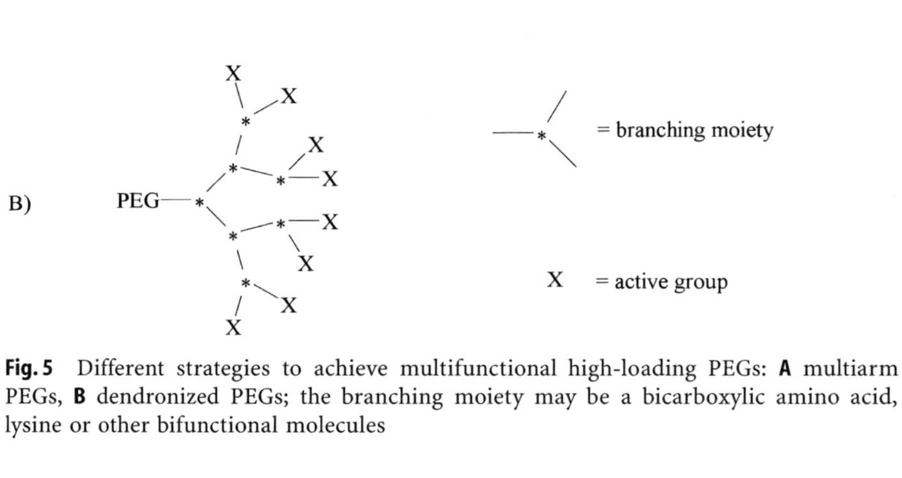

Fig. 5 Different strategies to achieve multifunctional high-loading PEGs: **A** multiarm PEGs, **B** dendronized PEGs; the branching moiety may be a bicarboxylic amino acid, lysine or other bifunctional molecules

triggers the linker (second prodrug), as reported for the drug delivery system based on a 1,6 elimination reaction or trimethyl lock lactonization [39, 40] (Scheme 1);
- Multiarm or "dendronized" PEGs (Fig. 5): the former are compounds prepared to link linear PEGs to a multimeric compound, whereas the latter are linear PEGs with a dentritic structure at one or both chain extremes [41, 42]. The aim of both derivatives is to increase the drug/polymer molar ratio, overcoming problems of high viscosity that may occur with mono-functional drug conjugate solutions. This is particularly true for a drug that requires a large amount of product for therapeutic treatments.

2.2
Forthcoming PEGs

Although more recent and selective PEGs have already been brought to market, some products—PEGs with new properties—may pave the way to further applications. The research on new PEGs may respond to different needs and can be summarized as follows: 1) the synthesis of monodisperse polymer, not only for low-molecular-weight PEGs (≤ 600 Da), which are already available, but also for PEGs higher molecular weights; 2) the design of PEGs with specific tailored groups to provide one side for linking to targeting molecules and the other side for linking followed by a controlled release of a therapeutic agent; 3) the preparation of high-loading PEGs that increases the drug payload—for example, by the construction of adendrimeric or multiarm structures at the level of the polymer extremes [41, 42]; 4) biodegradable PEGs that, thanks to a molecular weight reduction, can easily be cleared from the body.

3
Challenges in Protein PEGylation

PEGylation is a mature and well-documented technique that offers a wide selection of chemical methods for modification. However, unexpected difficulties may arise with any new protein, and several parameters have to be taken under consideration to achieve satisfactory results. In general, it is necessary to evaluate the effects of total linked PEG mass with respect to protein activity and specificity. One may remember, as an example, the in vivo advantages of PEGylated alpha-2b interferon with respect to the native protein, despite a great decrease in activity in an in vitro test following conjugation. In the design of a polymer–protein conjugate, it is also necessary to take into consideration both the PEG molecular weight and the number of attached polymer chains per protein. Moreover it is important to identify the best PEG derivative that will allow large-scale production, the feasibility of an adequate purification system, the maintenance of protein stability and activity, and, last but not the least, convenience in formulation.

Although each case of protein PEGylation requires its own discussion, the following guidelines may be useful in achieving good results:

1. An important parameter for the preparation of a new bioconjugate is the retention of the highest biological activity, although the advantages in terms of reduced immunogenicity and prolonged pharmacokinetic profile are equally and sometimes more relevant. For *enzymes*, the preservation of functionality is usually obtained when the linking of PEG chains does not disturb the active site or at least the residues involved in catalysis. Many strategies have been developed to achieve this goal: a) the use of branched PEGs that, due to their higher hindrance, may reduce the degree of modification in the active site (Fig. 4); b) to perform the PEGylation in the presence of a substrate or inhibitor that blocks the access to the active site by steric hindrance; c) to capture the enzyme on an insoluble resin, previously functionalized with an enzyme's substrate or inhibitor, and then conduct the PEG modification. In all of these cases the activated PEG is added to the complex, and the obtained conjugate is then eluted, for instance, by a change in pH or the addition of denaturants leading to a derivative where the active sites and its closer surroundings are free from PEGs (Fig. 6) [43]. For *other proteins*, an approach may consist of link-

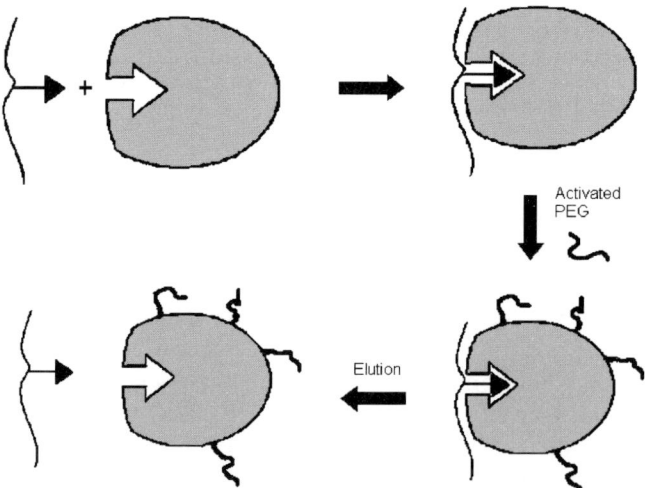

Fig. 6 Two-phase PEGylation strategy for the protection of an enzyme's active site from polymer conjugation: first, the enzyme is loaded into an affinity resin functionalized with an appropriate ligand. The enzyme's active site binds the ligand, thus protecting the active site itself and the area close to it from PEG modification. Finally, the modify enzyme is eluted from the column

ing the desired mass of PEG using a few high-molecular-weight polymer chains instead of a high number of low-molecular-weight chains. In fact, a multipoint attachment of PEGs on a protein surface usually reduces or prevents protein recognition via a shield effect, leading to loss in protein activity, an effect, that can be reduced linking one big PEG chain links to only one point of the protein. The effects of both number and mass of linked PEG chains on recognition and pharmacokinetic parameters are well documented in literature [44].

2. Direct-site PEGylation, as described above, may exploit the low pKa of the unique α amino group (pKa of 7.6–8.0); however, conjugation at the level of cysteine residue may offer further advantages. Cysteine is a rare amino acid that represents a specific point of attachment. Furthermore, if this amino acid is absent, it may be introduced at the desired position by genetic engineering to yield muteins with a cysteine residue replacing an amino acid not essential for the activity.

3. Proteins have several amino groups with different reactivity. Theoretically, using an excess of PEGs, all groups could be modified, and one may expect to obtain a homogenous product. However, many factors may interfere in polymer modification, such as a lack of accessibility and a non-deal pKa of amino groups, thus leading to mixtures of multi-PEG conjugates. When the amount of PEG is not in excess, a number of positional isomers are always formed. This last situation requires special attention to maintain reproducibility of the mixture over the preparation of different batches. Both differences, however, can be accepted by the authorities (FDA) as long as the identification of all adducts is provided [45, 46]. In this case, special skill is needed to fractionate the PEG isomer mixture. For this purpose, the possibility of exploiting different isoelectric points of the isomers by ionic exchange chromatography is very useful, while HPLC (high performance liquid chromatograpphy) reverse phase was found to be less efficient, and gel-filtration may separate only the species with different mass.

4. Once the isomers are separated, it is necessary to identify the localization of the PEGylation site in the primary amino acids sequence. The classical approach involves enzymatic digestion of the polymer–protein derivative, purification of the peptides, and the identification of these peptides by mass spectroscopy or amino acid analysis. A good example is reported in the characterization of PEGylated interferon α-2a [47]. Comparison of peptide fingerprinting of the conjugated protein with that of the native protein allows for an understanding of the region where PEGylation occurred based on the missing peptide. Besides the fact that this procedure is lengthy, however, the presence of the polymer may interfere with the analysis, since the cleavage by proteolytic enzymes can be incomplete due to steric hindrance. To circumvent this inconvenience, a new approach has been recently developed based on the use of tailor-made PEGs, PEG – Met – Nle – COOH or PEG – Met-βAla – COOH, which possess a chemically labile bond in the

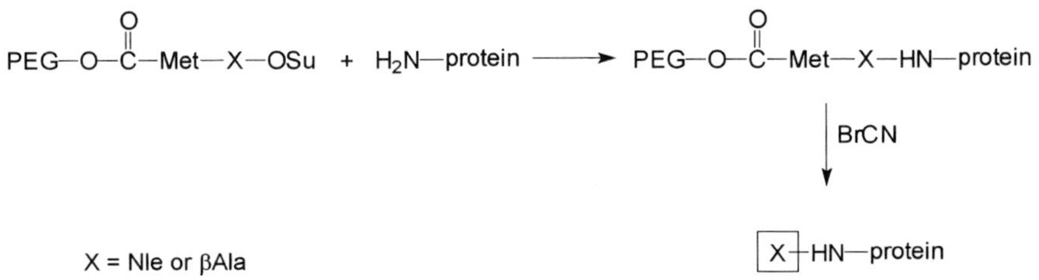

```
      O                                            O
      ‖                                            ‖
PEG—O—C—Met—X—OSu  +  H₂N—protein   ⟶    PEG—O—C—Met—X—HN—protein
                                                         │
                                                         │ BrCN
                                                         ▼

      X = Nle or βAla                          [X]—HN—protein
```

Fig. 7 Use of PEG – Met – Nle – OSu or PEG – Met-βAla – OSu to introduce a reporter amino acid at the PEGylation site: PEG conjugation followed by polymer moiety release by BrCN leaves Nle or βAla, which can be identified on the protein by amino acid sequence analysis

peptide spacer that can be cleaved by treatment with BrCN (Fig. 7). Following cleavage at the methionine level, the PEG chain is removed, and the remaining norleucine or β-alanine tags on the protein are identified by standard sequence investigation methods or by mass spectrometry analysis of the enzymatically digested fragments [48].

4
PEGylation Chemistry

In the past few years the chemistry of PEGylation was mainly focused on amino group modification and on developing new and original methods. This research resulted in an increase in yield and a selectivity in the binding site. Enzymatic approaches in obtaining PEGylated protein have been the most recent developments. This section briefly describes the strategies that have yielded successful products.

4.1
PEGylation on Protein Amino Groups

Amino groups are usually present on the surface of proteins and are easily accessible to reactive polymer molecules. That fact, together with the fact that lysine is a residue highly present in proteins, makes the amino group appealing for modification. Actually, all of the PEG-protein conjugates in the market come from an amino PEGylation conducted using different strategies. Also, for less reactive amino groups, such as the side chain of histidine, a PEGylation strategy has been reported and products have been developed [49]. A number of amino reactive PEGs, as reported in Fig. 2, have been synthesized and tested, differing in kinetic rate and in the resulting linkage (i.e., carbamate, amide, and secondary amine). A list of proteins PEGylated at the amino group is reported below with some comments on the chemistry employed and on the biological behavior.

4.1.1
Interferons

Type-1 interferons (IFN $-\alpha, -\beta, -\kappa, -\tau$, and $-\omega$) are cytokines often considered for PEGylation due to the drawbacks of the native proteins. Their activities include a wide range of effects—mainly antiviral, antitumor, and immunomodulation properties [50, 51]. IFN-α was first approved for hairy-cell leukemia therapy, later for treatment of hepatitis B and C, as well as for various dermatological pathologies, while INF-β was approved for treatment of multiple sclerosis. The low molecular weight of interferons ($\sim$ 20 kDa) is reflected in a relatively short serum half-life, a property that can be greatly improved by PEG conjugation.

One of the first studies involved the modification of interferon α-2a with linear succinimidyl carbonate PEG (SC-PEG; 5 kDa), in base buffer pH 10, via a urea linkage. The coupling, performed at an equimolar ratio of protein and polymer, led mainly to mono-PEGylated isomers and, in small amounts, to di-PEGylated conjugates and free interferon. Characterization of the conjugates indicated that lysine residues were the only site of PEGylation [47].

Reactions carried out in phosphate buffer at pH 6.5 demonstrated that derivatives with improved pharmacokinetic profiles and higher activity could be obtained by the conjugation of interferon α-2b with SC-PEG (12 kDa). This reaction gave an unexpected conjugate at histidine-34, representing approximately 47% of the total PEGylated species (Scheme 2) [49]. The stability of this conjugate was studied by ^{1}H-NMR analysis following the chemical shift of $H^{\varepsilon 1}$ in His-34 [52]. The activity of this interferon preparation was related to its ability to release free and fully active interferon by slow hydrolysis of the labile His-PEG bond [53]. These studies paved way for PEG-Intron® to hit the market in 2000. Although the in vitro potency of this PEG-interferon is only 1/4 of the free form, its serum residence time is about six times longer, allowing for a less frequent administration schedule while maintaining an efficacy comparable to unmodified interferon [18, 54].

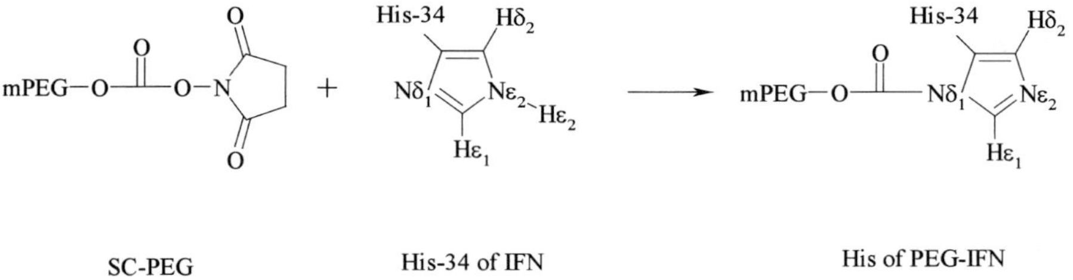

Scheme 2 Adduct formation at the level of His-34 in interferon α-2b using SC-PEG as a PEGylating agent

A different approach to interferon PEGylation exploited the special properties of branched PEGs. A high-molecular-weight branched PEG (PEG$_2$, 40 kDa) was chosen on the basis of several preliminary studies revealing that: a) the protein surface protection with a single, long and hindered chain PEG is higher than the one obtained with several small PEG chains linked at different sites [8]; b) branched PEGs have lower distribution volumes than linear PEGs of identical molecular weight, and the delivery to organs such as the liver and spleen is faster [55]; c) proteins modified with branched PEGs possess greater stability with respect to enzymes and pH degradation [34]. The 40 kDa branched succinimidyl PEG (PEG$_2$-NHS) was linked to interferon α-2a using a 3 : 1 PEG/protein molar ratio in 50 mM sodium borate buffer pH 9 (Scheme 3) [19]. PEGylation under these conditions led to a mixture containing 45–50% mono-substituted protein, 5–10% poly-substituted protein (essentially dimer), and 40–50% unmodified interferon (Fig. 8). Identification of the major positional isomer within the mono-PEGylated fraction was carried out by a combination of high-performance cation exchange chromatography, peptide mapping, amino acid sequencing, and mass-spectroscopy analysis. It was demonstrated that PEG was attached mainly to either Lys-31, Lys-121, Lys-131, or Lys-134 [19]. Even though the in vitro antiviral activity for PEG2-IFN was greatly reduced (only 7% of residual activity was found), the in vivo activity, measured as the ability to reduce the size of various human tumors, was higher than that of free IFN. The positive result could be related to the extended blood residence time of the conjugated form as shown in Table 2. These studies brought into the market a long-lasting blood interferon conjugate, Pegasys®, which is effective in eradicating hepatic and extrahepatic hepatitis C virus (HCV) infections [20].

Table 2 Pharmacokinetic properties of interferon α-2a and its PEGylated form in rats [20]

Protein	Half-life (h)	Plasma residence time (h)
interferon α-2a	2.1	1.0
PEG2 (40 kDa)-interferon α-2a	15.0	20.0

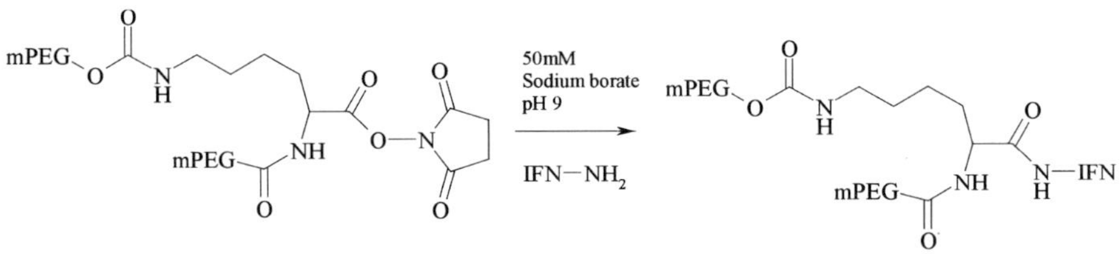

Scheme 3 PEGylation of interferon α-2a by branched mPEG$_2$ – COOH (40 kDa)

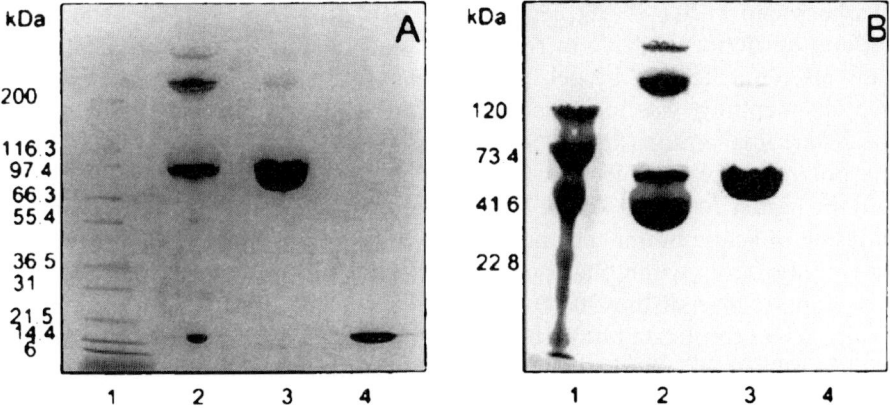

Fig. 8 SDS-PAGE analysis of the PEGylated interferon α-2a mixture. The conjugates were: **A** specifically stained for protein with Coomassie blue—lane 1: molecular weight marker proteins; lane 2: PEGylation reaction mixture; lane 3: purified PEG_2-IFN; and lane 4: interferon-2a. **B** Specifically stained for PEG with iodine—lane 1: molecular weight marker PEGs; lanes 2–4: same as in Fig. 2A. Note that lane 4, containing interferon α-2a in gel B, is not stained by iodine, reproduced from [19]

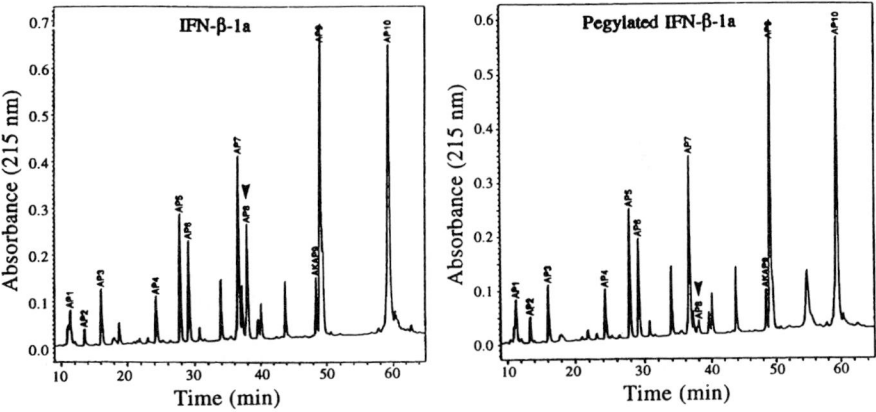

Fig. 9 Localization of the site of PEGylation by C_4-HPLC peptide mapping following digestion with endoproteinase Lys-C to release peptide 1–19. *Left*: unmodified IFN-β-1a. *Right*: PEGylated IFN-β-1a. *Arrowheads* (pointing to peak AP8) mark the elution position of the peptide that has disappeared in the PEGylated protein, *right panel*. Reproduced from [55]

A different cytokine, IFN-β is approved for the treatment of multiple sclerosis in the United States, but like other cytokines, it suffers from a short blood residence, again suggesting PEGylation as a solution. An exhaustive study conducted by Pepinsky et al. [55] resulted in a PEG modification of interferon-β-1a that exploited a reductive alkylation of an amine residue in phosphate buffer at pH 6 with an excess of 20 kDa PEG aldehyde and sodium cyanoborohydride to reduce the intermediate Shiff base. The conjugate, after purification by gel-filtration, ion exchange chromatography, and peptide mapping, was identified as mono-PEGylated interferon modified at the N-terminal amine (Fig. 9). Such a result could be achieved thanks to both the polymer's high molecular mass, which prevents multiple conjugations, and the higher reactivity of the N-terminal amine with respect to the ε-amine of lysine in acidic buffer. The modified protein retained the same potency as native interferon, while pharmacokinetic experiments showed a five-fold increase in serum half-life. 20 kDa PEG was chosen since the 5 kDa PEG led to a fully active conjugate but with no improvement in pharmacokinetic profile, and the 40 kDa PEG derivative was completely inactive.

4.1.2
Granulocyte Colony Stimulating Factor (G-CSF)

Granulocyte colony stimulating factor (G-CSF) is a cytokine that controls proliferation, differentiation and functional activation of neutrophilic granulocytes [56]; it is already used to treat granulocyte depletion during chemotherapy [57]. A number of PEGylation studies were also carried out on G-CSF muteins, a class of genetic variants with higher activity with respect to the native human protein. Several PEGs with different molecular weights and active moieties were investigated, and the best one was the 20 kDa mPEG-succinimidyl propionic acid (mPEG-SPA). The in vitro activity of these PEG-G-CSF conjugates decreased with an increase of either the number or the molecular weight of PEG chains linked to the protein, while the in vivo residence time was directly proportional to the total mass of the PEG [58, 59]. The advantages due to the extended blood residence time of conjugates were found to overcome the lower affinity toward the G-CSF receptor. These findings indicate once again the difficulty of evaluating the success of a polymer–protein conjugate on the basis of in vitro assays only.

An interesting PEG modification was recently proposed for a multimeric single chain G-CSF [60]: a covalent dimer obtained by a DNA recombinant technique using an expression vector encoding for two G-CSF genes in series. The dimeric protein was also produced with one or more mutated sites in the amino acid sequence to improve the biological activity and allow a more specific PEGylation. PEG-SPA of different MWs (preferably 5, 12, and 20 kDa) was used in the modification, unfortunately leading to conjugates with lower activity as compared to native G-CSF.

Kinstler proposed a reductive alkylation strategy with PEG-aldehyde/ sodium cyanoborohydride in acidic buffer solution (Scheme 4) to modify selectively the low-pKa amino groups [22]. Under these conditions, modification limited to the N-terminal methionine of r-metHuG-CSF was obtained (Fig. 10) [61]. The selective modification of the α-amino group with respect

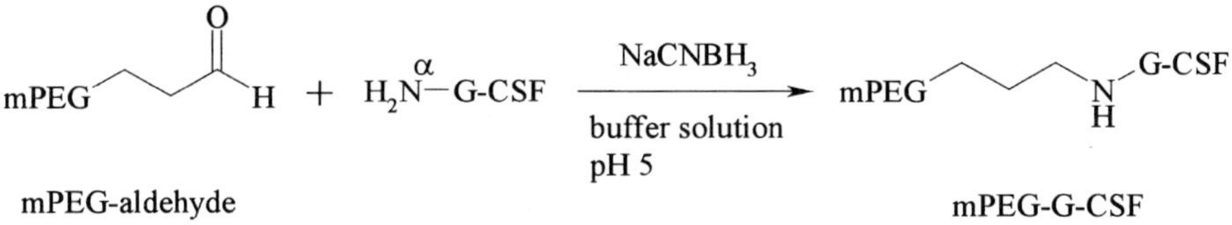

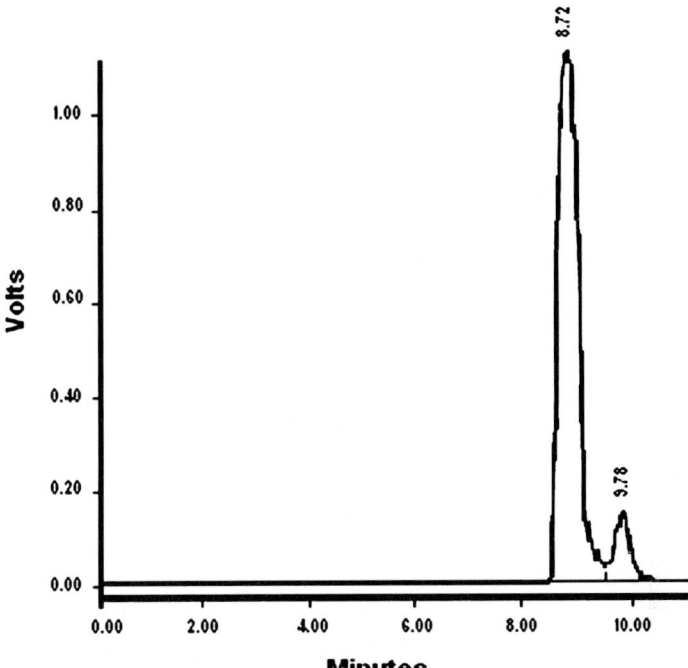

Scheme 4 Mono-mPEG-G-CSF conjugates were prepared by reductive alkylation of the α-amino group of the N-terminal methionine residue of r-metHuG-CSF with mPEG-aldehyde

Fig. 10 Size-exclusion HPLC analysis (UV detector at 280 nm) of the reduced mixture of r-metHuG-CSF reacted with mPEG aldehyde (M_w = 6 kDa): N-terminal mono-mPEG–G–CSF conjugate eluted at 8.72 min (92% of total area); unreacted r-metHuG-CSF eluted at 9.78 min (8% of total area). Reproduced from [22]

to the ε-amino residue of lysine is explained on the basis of the differences in pKa (7.6–8.0 and 9.3–9.5 for α and ε respectively): moderately acidic pH values leave the α-amino group still reactive [62]. The conjugate with a molecule of PEG 20 kDa shows an improved pharmacokinetic profile mainly due to reduced kidney excretion. In addition, the clearance of G-CSF is also due to an internalization process of the receptor–ligand complex in neutrophile cells, which it is in some way related to the number of circulating neutrophileses. The PEG-G-CSF, which remains in the bloodstream for a prolonged time, stimulates the proliferation of neutrophiles, and consequently, under therapy, causes its own clearance. The PEG-G-CSF conjugate, Pegfilgastrim®, has been on the market since 2002.

4.1.3
Megakaryocyte Growth and Development Factors (MGDFs)

Thrombopoietin or megakaryocyte growth and development factor (MGDF) is a key regulator of thrombopoiesis [63] that acts on the expansion and maturation of megakaryocyte progenitor cells, resulting in increased platelet counts [64]. The level of thrombopoietin is thought to be self-regulated, mainly by internalization, after binding to the cell surface receptors present in megakaryocytes and platelets [65].

A truncated form of human MGDF, comprising the 1–163 sequence of the native protein (rHuMGDF) exhibiting a specific activity five-fold higher than the full-length hormone, was PEGylated with the same chemical strategy already proposed in G-CSF modification [61, 66]. The truncated form is devoid of the glycosyl moiety present in the native protein. In this case, PEGylation also prevented the specific non-enzymatic degradation of the protein; in particular, the chemical inactivation of MGDF due to cyclization and cleavage of the first two amino acids, leading to diketopiperazine and des(Ser,Pro)rHuMGDF (Scheme 5). Such a reaction could be avoided by PEGylating the *N*-terminal amine with PEG aldehyde [67]. When tested in platelet-count experiments, the PEG derivative was as active as the glycosylated full-length native thrombopoietin and much more active than both the free rHuMGDF and the non-glycosylated full protein [22]. The truncated form of the protein was selected for PEGylation instead of the native one, yielding a conjugate with comparable activity and improved pharmacokinetics with respect to the full-length hormone, demonstrating that PEGylation may replace glycosylation. This finding could open doors to new strategies and the potential for protein PEGylation.

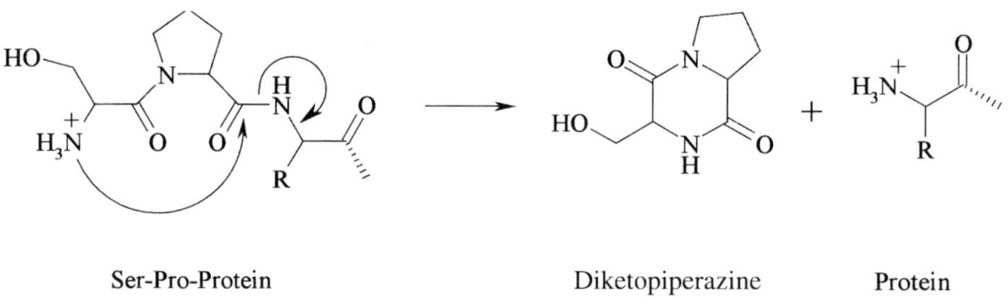

Ser-Pro-Protein   Diketopiperazine   Protein

Scheme 5 Chemical inactivation of rHuMGDF by cyclization and cleavage of the first two *N*-terminal amino acids, Ser and Pro. The degradation leads to the formation of diketopiperazine and des(Ser,Pro)rHuMGDF

4.1.4
Growth Hormone (GH), Growth Hormone-releasing Hormone (GRF), and Growth Hormone Antagonist

Human growth hormone (hGH) is a protein that has several effects, including linear body growth, tissue growth, activation of macrophages, lactation, and insulin-like and diabetogenic effects [68]. It carries out its action through receptor recognition and signaling, which requires receptor dimerization. Clinically it is used to treat GH deficiency, which causes short stature, and both Turner's syndrome and cachexia in AIDS patients. A long-lasting form would be welcome, since, for its short in vivo half-life, hGH must be administered on a daily basis. One of the first studies to this end involved a random PEGylation of hGH with low-molecular-weight PEG-N-hydroxysuccinimide (PEG-NHS 5 kDa) leading to a mixture of derivatives. Upon purification, hGH conjugates with up to seven PEG chains per hGH molecule (Fig. 11) could be

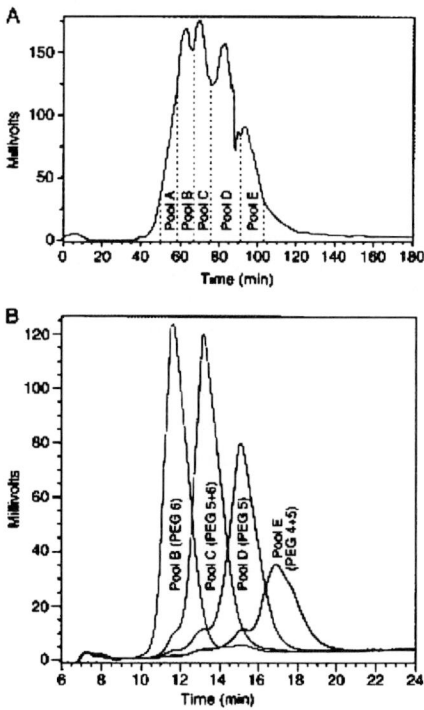

Fig. 11 Preparative SP-Sepharose high-performance chromatography of a PEG_{5000}-hGH reaction mixture (*panel A*). Fractions were pooled as shown on the chromatogram and analyzed by mass spectrometry to determine the average number of PEG_{5000} groups attached per hGH. Purity for four of the five peaks was further assessed by analytical high-performance liquid chromatography on a sulfopropyl TSK 5PW column (*panel B*). Reproduced from [69]

isolated [69]. Binding studies demonstrated an inverse correlation between the number of PEG chains and receptor affinity, as PEG reduced the association rate for receptor/adduct formation. By contrast, the higher the number of PEG chains linked, the higher the in vivo potency, due to reduced kidney clearance. It was found that hGH conjugated to five PEG chains was a good compromise for therapeutic purposes, presenting a ten-fold higher in vivo potency than the unmodified hormone.

An alternative treatment for GH deficiency involves the use of growth hormone-releasing hormone (GHRH) [70] or its analogs, commonly reported here as GRF. With its 44-amino-acid sequence, GHRH controls the levels of expression and release of GH, but it is also naturally present in truncated active forms comprising only 40 or 37 amino acids. Furthermore, the 1–29 truncated peptide of GRF (hGRF$_{1-29}$), commonly prepared by chemical synthesis, retains the same activity, in vitro and in vivo, as the full-length hormone [71]. However, the clinical use of this truncated form of GRF is severely limited by its short biological half-live (10–20 min in humans), both due to fast kidney excretion and due to N-terminal enzymatic degradation by endogenous aminopeptidases. PEGylation was therefore considered, since it could act on both undesired effects. Esposito et al. [44] coupled PEG-NHS (5 to 20 kDa) to hGRF$_{1-29}$, a peptide that has three available conjugation sites (Lys-12, Lys-21, and the N-terminal α amine). PEG-NHS of molecular weight up to 20 kDa was used in the modification, and DMSO was chosen as solvent, since, beyond offering major stability and solubility to GRF than the aqueous media, it could induce a conformation of the peptide that lowers the amount of PEGylated isomers. The activity of PEG conjugates was found to be dependent on the total mass of linked PEG: the higher the PEG mass, measured either as the number of polymer chains or as molecular weight, the lower the potency (Table 3). It was found that, under suitable reaction conditions and after chromatographic purification, PEG$_{5000}$–OSu yielded a solution with prevalent amounts of mono-PEGylated Lys-12 and Lys-21 isomers in an equimolar mixture (Fig. 12) [72]. The mono-PEGylated mixture was purified, from unmod-

Table 3 Influence of total PEG mass on GRF conjugate activity, as assessed by a specific reporter gene assay [44]

Products	EC$_{50}$
Native GRF	0.18
GRF-PEG 5000	1.06
GRF-PEG 10 000	2.8
GRF-PEG 20 000	> 1000
GRF-(PEG 5000)$_2$	31.8
GRF-(PEG 5000)$_3$	79.6

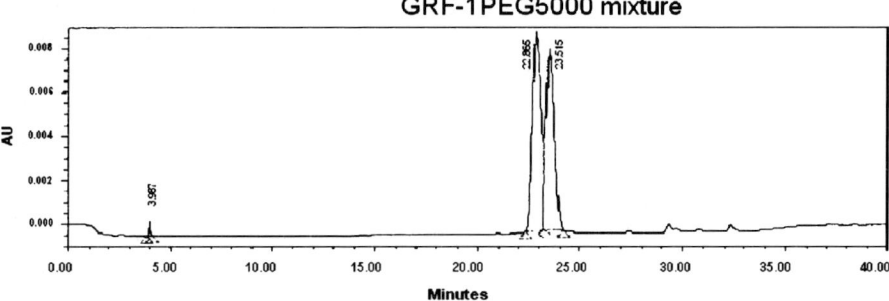

Fig. 12 RP-HPLC C_8 analytical chromatogram of the crude reaction mixture in DMSO of GRF before purification on the ion-exchange purification column. Reproduced from [72]

ified GRF and di-PEGylated conjugate, by ionic exchange chromatography and characterized by MALDI-TOF/MS analysis (Fig. 13) [44]. The conjugates showed an improved pharmacokinetic profile, which finally resulted in a more favorable pharmacodynamic response on the growth hormone-insulin grow factor 1 (GH-IGF-1) axis, as demonstrated by increased GH levels and the number of peaks in pig and rat plasma [73]. This work pointed out the importance of choosing the right solvent for the coupling reaction, which may reduce the number of PEG isomers by promoting structural rearrangement of the peptide. This could consequently lead to easier purification and higher yields. In fact, investigation conducted by NMR and circular dichroism in-

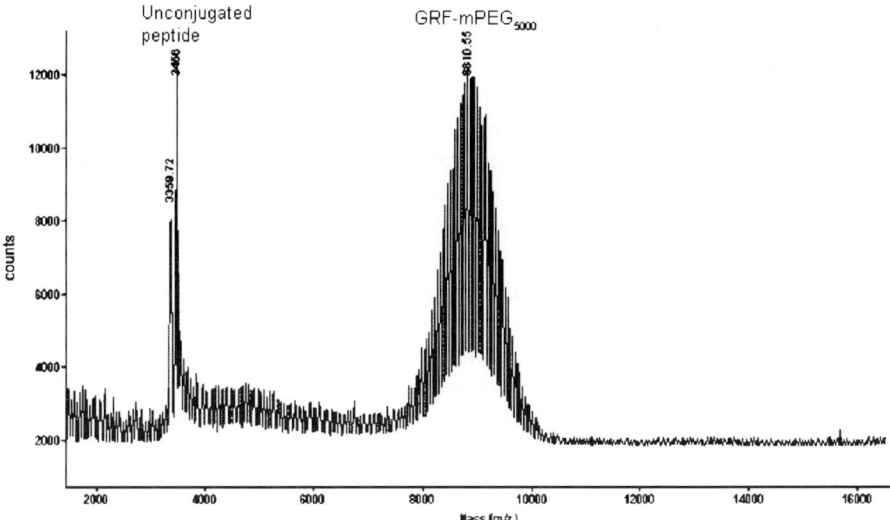

Fig. 13 MALDI-TOF/MS investigation of GRF_{1-29} mono-PEGylated mixture. Signals at 8810 correspond to mono-PEGylated GRF. The peak shows the typical polydispersivity of PEG. Minor signals for unmodified peptide were also detected. Reproduced from [44]

dicated that the α-helix percentage in hGRF$_{1-29}$, which is only 20% in water, is raised to 90% in structure-promoting solvents such as methanol/water or 2,2,2-trifluoroethanol. In the latter solvent, the percentages of PEGylation at the level of Lys-12 reached 80% of the total amount of PEGylated isomers with different PEGylating agents [74]. This suggested that a highly regio-selective PEGylation could be achieved in the proper solvent.

Acromegaly is a disease caused by an over-expression of hGH. The therapy consists of the administration of a recombinant hGH mutein (B2036) with an antagonist effect. GH can bind two chains of its receptor thanks to two different binding sites, while B2036 can interact with only one receptor chain. B2036 is obtained through mutations of hGH at both binding sites with the aim of retaining the binding capacity of site 1 and blocking the activity of site 2, thus avoiding receptor signaling [75]. Furthermore, in view of PEG conjugation, a mutation was smartly performed to direct PEGylation to site 2 only. To reach this goal, a lysine residue at site 1 was exchanged with a different, non-reactive amino acid, while a lysine was added to site 2. SPA-PEG (5 kDa) was employed for PEGylation, forming a mixture of derivatives with 4 to 6 PEG chains for each B2036. Further studies on the conjugate revealed a reduced binding affinity for the cellular receptor that was counterbalanced by a prolonged blood residence time [76]. This result may encourage the modification of proteins where the PEG moieties, through steric entanglement, may enhance or decrease the protein binding affinity and change the final protein behavior.

4.1.5
Antibodies and Antibody Fragments

An increasing interest has been focused on antibodies (see [77] for an overview) for the treatment of several diseases. Ten products based on these entities have already reached the market [78], and more are undergoing clinical evaluation [79]. PEGylation was initially applied to these proteins to reduce the high immunogenicity of murine monoclonal antibodies. Humanized and human antibodies seemed to be the solution to the problem, but the high production costs and the low level of expression limited their use to severe diseases. Recently, an *Escherichia Coli* expression of antibody fragments, such as Fab' and scFv (Fig. 14), was proposed, allowing a cutting in the cost of their production and, consequently, a potential increase in the market. However, since their major limitation is their short half-life, PEGylation was widely investigated. Many studies have been reported on random PEGylation (the reader is encouraged to refer to the extended review by Chapman [80]), but almost all of the conjugates showed a significant reduction in binding affinity. PEGylation of anti-IL8 F(ab')$_2$ was an exception, since the coupling with PEG of 20 or 40 kDa did not heavily affect the binding affinity [81].

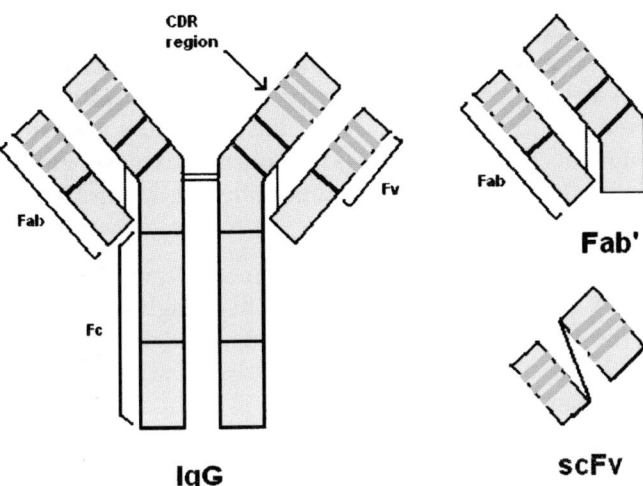

Fig. 14 Schematic drawing of an IgG antibody and its fragments

The reduced binding affinity of PEGylated antibodies is either due to a modification in proximity of the antigen binding domain, generating steric hindrance to the binding, or to a reduction of the number of free lysine residues involved in the ionic interactions that initiate the antigen binding. For example, a limited modification of the Fc region caused a dramatic loss in binding activity. Fc (see Fig. 14) is the region that mediates the effector functions of the antibodies, either by complement activation or by cellular interaction after interaction with its Fc receptors. The high number of lysines in the Fc domain drives the PEGylation to this region so that, when the degree of lysine modification is greater than 15%, the binding activity of the Fc receptor is completely impaired [82]. PEGylation of antibodies, as a means to reduce their immunogenicity, has been overcome by technical advances in the humanization of murine antibodies or by expression of human antibodies in bacteria. Nonetheless, PEGylation still remains important for antibodies used in cancer therapy, because of the positive enhancing effect of PEG on antibody tumor localization and the lack of Fc effector functions.

Engineered antibody fragments represent a major advance in the design of a shorter polypeptide sequence retaining binding activity towards the antigen. Such polypeptides are called single-chain Fv fragments (scFV, see Fig. 14). The major limitation of these peptides is, however, a half-life that is too short (about 0.7 hours), which prevents their use in any therapeutic application. Enzon researchers have synthesized PEGylated scFV conjugates using several PEG derivatives (linear and branched SC-PEG and pNCP-branched PEG, among others). These conjugates possess improved pharmacokinetics and lower immunogenic activity than a full-length antibody, since, after PEGylation, the peptide is a small fraction of the total molecule.

As mentioned above, PEGylated antibodies and antibody fragments are still of great interest and are widely studied, mainly in oncology. These conjugates are used as targeted delivery systems for anticancer drugs, since several studies showed that PEGylation increases the accumulation of antibodies and antibody fragments in a tumor, but not in normal tissues. The advantages of these PEGylated derivatives over the unmodified antibodies may be summarized as follows:

- high tumor accumulation due to the enhanced permeability and retention effect (EPR), because most solid tumors have high vascular permeability allowing the extravasation of high-molecular-weight molecules that accumulate in the tissue due also to inefficient lymphatic drainage [83];
- increased plasma half-life;
- decreased liver uptake [84].

The antibody fragments, with respect to full-length antibodies, possess higher tumor penetration potential [85], making them promising candidates for tumor-targeted delivery systems. The strategy was demonstrated by coupling the C225 monoclonal antibody, directed against an epidermal growth factor receptor, to a heterobifunctional PEG having a radiometal chelator (diethylenetriaminepentaacetic acid, DTPA) at one end [86]. The conjugate DTPA-PEG-C225, with up to 60% of the C225 amino groups modified, retained 66% of its binding affinity, and, more importantly, when labeled with Indium-111 (^{111}In), it demonstrated a narrower steady-state distribution than the non-PEGylated ^{111}In-DTPA-C225 due to reduced non-specific binding. Another example is represented by derivatives where the anticancer drug is linked to a PEG chain conjugated to the monoclonal A33 antibody leading to a conjugate useful for colorectal carcinoma treatment [87]. The A33 antigen is expressed in high amounts in some colon cancer cell lines, such as SW1222. Moreover, it was also reported that PEG-modified Technetium-99m-radiolabeled antibody fragments was useful for radioimmunodetection of tumors and infectious lesions [88].

4.1.6
Others Proteins

Insulin is one of the most studied small proteins for PEGylation, and its modification has been carried out through both random and site-specific methods. The protein is formed by two polypeptide chains, A and B, and its three amino groups (Gly-A1, Phe-B1, and Lys-B29) are candidates for PEGylation. Hinds, among others, proposed a site-directed PEGylation procedure involving the preliminary preparation of insulin protected by N-BOC (tert-butyl chloroformate) [89]. As an example, to synthesize $N^{\alpha B1}$-PEG-insulin, the intermediate $N^{\alpha A1}$, $N^{\varepsilon B29}$-BOC-protected insulin was prepared before conjugation with PEG-SPA (MW 750 or 2000 Da). After the removal of the t-Butyloxy carbonyl group (BOC) by acid treatment with trifluoroacetic acid (TFA), the resulting conjugates $N^{\alpha B1}$-PEG$_{750}$-insulin and $N^{\alpha B1}$-PEG$_{2000}$-insulin retained 104% and 83% of the native insulin bioactivity, respectively. The Lys-B29 PEG conjugates, obtained using the same procedure, led to comparable results in term of bioactivity. Furthermore, all derivatives showed a reduced self-aggregation and extended half-life with respect to the unmodified insulin. Similarly, the acylation of Gly-A1 and Lys-B29 amino groups with the cyclic anhydride of a bicarboxylic acid, protects these amino acids from the following PEGylation at the level of free Phe-B1. Deprotection of Gly-A1 and Lys-B29 has been performed under mild acid conditions, thus leaving only the PEG chain linked to the Phe-B1.

Arginine deaminase from microbial origin has been studied as potential anticancer enzyme, since it degrades arginine, an essential nutrient for some tumors; however, its use is limited due to short half-life and high immunogenicity. The enzyme was modified with PEGs of various molecular weights and structures (branched and linear), and with different coupling chemical strategies [90]. It was demonstrated that 50% of the enzyme activity was maintained even when up to 40% of amino group were modified. With a PEG MW of up to 20 kDa, linearity was found between PEG weight, pharmacokinetic, and pharmacodynamic properties; in fact, conjugates with a prolonged blood residence time, which was obtained using higher-MW PEGs, showed a better pharmacodynamic response. Reduced immunogenicity was also found for the PEGylated microbial enzymes. Arginine deiminase was additionally modified with SC-PEG (12 kDa) to obtain conjugates with 16 to 22 polymers chains for each enzyme subunit and a 20% residual activity. Engineered arginine deiminases were designed to allow PEGylation far from the catalytic region—in this case, the conjugation with succinimidyl succinate PEG (SS-PEG) yielded a product with 70 to 80% of native enzyme activity.

Erythropoietin (EPO) is a glycoprotein that increases the production of reticulocytes and red blood cells by the stimulation of bone marrow cells. Recombinant EPOs (rEPOs) has an increased number of glycosylation sites, and mono-PEGylated conjugates of the new rEPOs showed in vivo potency higher than the unmodified rEPOs [91]. The PEG conjugates are administered once a week instead of three times a week like the native form. A further study proposed PEGylation of EPO at the Lys-54 residue, through optimization of the polymer/EPO ratio and the conjugation reaction time. This controlled reaction led to a mixture of EPO conjugates with one to three PEG chains, and the desired mono-PEGylated EPO at Lys-54 was isolated.

Leptin (OB proteins) is secreted from adipose tissue and plays a role in body weight homeostasis by regulating food intake and energy expense, and was suggested in the treatment of obesity. An interesting modification method of OB involves a dual PEGylation of a recombinant OB (rOB) where Arg-78 is mutated into Cys-78. The linking of two PEG chains with different reactivity in two different protein sites is therefore possible. First, MAL-PEG

was coupled to Cys-78, and later, PEG-aldehyde was reacted with the protein's
N-terminus at an acidic pH [92]. The conjugate was found to possess higher
activity and biocompatibility compared with the unmodified OB. A modification of OB protein with branched PEG2-NHS was also carried out, and the
purified mono-PEGylated form was tested in mice for its ability to reduce
food intake and body weight [93]. MonoPEG2-OB was found active over three
days after injection, proving an extended blood residence time.

4.2
PEGylation on the Protein Thiol Groups

Since it occurs rarely in proteins, a free cysteine residue represents an optimal
situation to achieve direct-site modification. PEG derivatives having specific reactivity toward thiol groups have been developed—namely, MAL-PEG,
OPSS-PEG, PEG-iodoacetamide (IA-PEG), and VS-PEG (Fig. 15). PEGylation
at the level of cysteine makes both purification and characterization of conjugates easier due to the absence of either positional isomers or products with
different degrees of substitution. The potential of cysteine PEGylation can be
also exploited when cysteine is not naturally present in the protein, since genetic engineering techniques may introduce this amino acid in the sequence
by insertion at desired positions or by substituting cysteine for a suitable
amino acid. Examples of site-specific PEGylation through the chemistry of
cysteine thiol modification is reported below.

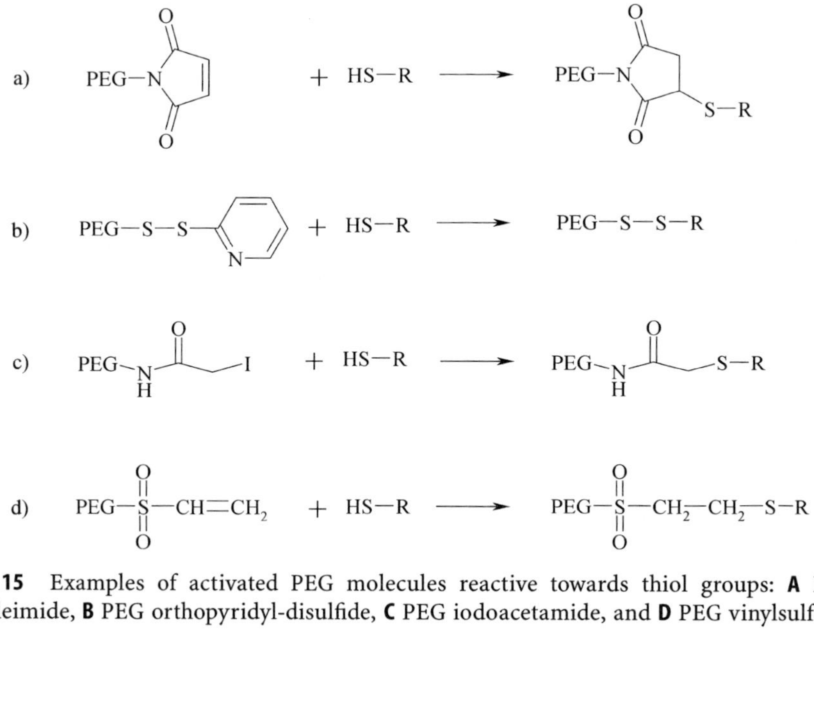

Fig. 15 Examples of activated PEG molecules reactive towards thiol groups: **A** PEG maleimide, **B** PEG orthopyridyl-disulfide, **C** PEG iodoacetamide, and **D** PEG vinylsulfone

IFN-β was PEGylated at the level of cysteine 17 [94] using as thiol-reactive agent the PEGylating agent OPSS-PEG. Conjugation could be specifically directed to cysteine 17, which is the only free thiol group present. Derivatization was carried out with PEGs of different molecular weights, but with a special and unique strategy to overcome the problem of low yield obtained with high-molecular-weight polymers. Low-molecular-weight PEGs are more reactive and overcome the steric hindrance around the thiol group in this protein. On the basis of this observation, modification with high-molecular-weight PEGs was obtained via a two-step procedure: in the first one, the protein was modified with a low-molecular-weight, heterobifunctional PEG oligomer, and in the second step by conjugation with a higher-molecular-weight PEG possessing specific reactivity towards the free terminal end of the first oligomer (Fig. 16). This strategy implied the use of a heterobifunctional PEG oligomer with a thiol-reactive group at one extreme and a hydrazine group at the other. Hydrazine was chosen since it is characterized by a very low pKa (3–4) as compared to the protein amine residue pKa values (pKa 7–9.5). The PEG oligomer selected for this conjugation was therefore the OPSS-PEG-hydrazine (OPSS-PEG-Hz, 2 kDa) leading, in the first step, to INF-SS-PEG-Hz, which could be selectively modified with PEG-aldehyde (30 kDa) by reductive alkylation. The overall yield was higher than 80%, and PEG-INF-β-1a from Serono has now completed phase I of clinical trials.

The hGH protein was extensively studied for PEGylation and, to overcome poly-substitution, cysteine muteins were synthesized by a recombinant DNA technique. Among all of the possible mutations described in the literature and patents, the cysteine addition at the C-terminus of hGH allows one to obtain a mutein that retains full activity and allows site-specific PEGylation with a cysteine-reactive PEG-maleimide (PEG-MAL, 8 kDa). It was necessary

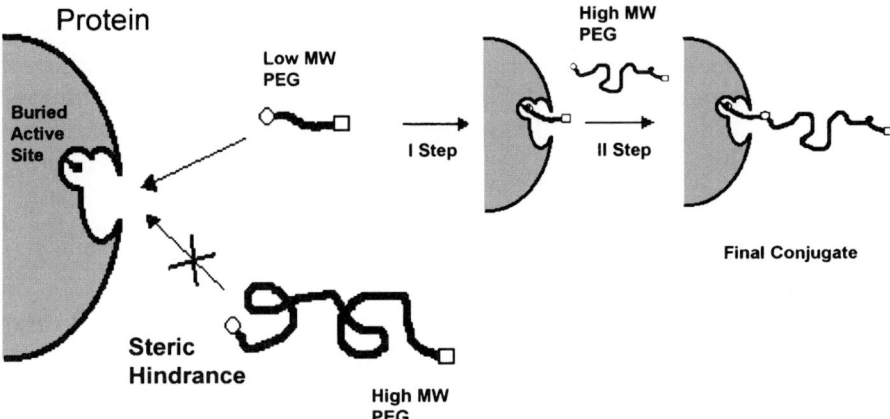

Fig. 16 Two-step tagging PEGylation strategy for a buried SH group in protein. Smaller PEG molecules are more reactive than high-molecular-weight PEGs

to treat the rhGH mutein with an excess of 1,4-dithio-DL-threitol (DTT) before the coupling step, to convert this new carboxyl terminal cysteine in the reduced form and prevent its inclusion in the formation of disulphide bridges. After the removal of an excess of DTT by gel-filtration, the conjugation resulted in a mono-PEGylated derivative with over 80% yield [95].

Interleukin-2 (IL2) is a powerful immunoregulatory lymphokine [96] produced by lectin- or antigen-activated T cells that, like interferons, enhances the natural killer cell activity and therefore may find a role in the treatment of cancer. There is evidence that IL2 can act as a growth hormone for both B and T lymphocytes. Goodson and collaborators [97] proposed an interesting modification strategy for recombinant IL-2, where the native glycosylation site was substituted by a PEG chain. To achieve this goal, they introduced a cysteine residue at the glycosylation site, thus obtaining 3Cys-rIL-2, which was later conjugated to PEG-MAL. The derivative retained the full activity of the parent protein and had a four-fold increase in blood residence time. In this case, PEG-3Cys-rIL-2 mimics native IL-2, with the PEG chain replacing the sugar moiety. This method may better preserve the activity since glycosylation regions, due to the steric bulk of the sugar moiety, are commonly not involved in receptor binding.

An alternative approach in *antibody* PEGylation, which does not involve amino modification, exploits the use of engineered antibody fragments, such as Fab' (see Fig. 16), with at least one free cysteine residue located far from the antigen binding site and reacting specifically with PEG-maleimide, leading to mono-PEGylated forms. Studies based on this strategy and employing a PEG molecular weight of up to 40 kDa demonstrated that the binding affinity was completely retained for all PEG-engineered Fab' conjugates, as opposed to the random PEGylation of Fab' NH_2 groups that led to up to a 50% loss of binding activity [98].

Hemoglobin (Hb) has been extensively investigated as an oxygen-carrying therapeutic agent, but the limitations for its clinical use come from its high vasoactivity due to extravasation into interstitial spaces and the subsequent scavenger action on nitric oxide. PEGylation has been performed to prevent Hb extravasation. After several unsuccessful attempts through random PEGy-

Table 4 Molecular parameters of PEGylated hemoglobin [25]

Hemoglobins	Calculated mass (kDa)	Radius (nm)	Volume (nm^3)
Hb	64	3.12	127
$(PEG5kDa)_2$-Hb	74	4.20	310
$(PEG10kDa)_2$-Hb	84	5.54	712
$(PEG20kDa)_2$-Hb	104	7.04	1436
Hb-octamer	128	4.12	293
Hb-dodecamer	192	5.56	720

lation, a site-specific modification was performed at Cys-93(β) with maleimidophenyl PEG (Mal-Phe-PEG) (5, 10, and 20 kDa), leading to PEGylated Hb carrying two chains of the polymer per Hb tetramer [99]. The colligative properties of the derivatives suggested that PEG helped to eliminate the Hb vasoactivity [24]. This product was found to be more efficient than polymerized Hb, Hb-octamer, or the dodecamer. PEG_2-Hb conjugates also possess a higher oxygen affinity, independently of the polymer chain length. A very detailed study [25] demonstrated that the hydrodynamic volume and the molecular radius of the conjugates increase linearly with an increase of PEG molecular weight (Table 4), while the viscosity and the colloidal osmotic pressure exhibit an exponential correlation with the same parameter. The authors concluded that the vasoactivity modulation is due to the surface shielding effect of PEG.

4.3
Protein PEGylation Catalyzed by Enzymes

A PEGylation procedure that differs from the usual chemical methods involves the use of enzymes that catalyze the conjugation of polymer chains to a specific site on a protein surface. Since the first studies in this direction, proposed by H. Sato, involving the enzyme transglutaminase (TGase) [100], many other researchers have proposed interesting approaches using different enzymes. The enzyme-catalyzed coupling seems to yield more homogenous derivatives than chemical PEGylation, known to result in a mixture of isomers. This innovation seems to open doors to new derivatizations, since the selectivity and the wide selection of enzymes may enable polymer linking to specific protein amino acids in the sequence or to residues that may not be reached by the usual PEGylation chemistry.

Sato studied two PEGylation methods for IL-2 using two different transglutaminases (TGase), one coming from guinea-pig liver (G-TGase) and the other one from micro-organism *Streptoverticillium sp.* strain s-8112 (M-TGase). Both enzymes catalyze the transfer of an amino group from a donor (for example, PEG-NH_2) to a glutamine residue present in a protein (Scheme 6). The difference between the two enzymes involves the requirement of a transfer to take place in the amino acid sequence. G-TGase activity is more strictly dependent upon the amino acids surrounding the glutamine. For PEGylation, several PEGs of different molecular weight, terminating with a linear chain alkylamine at one end, were synthesized. Although IL-2 contains six glutamines, none of them is a suitable substrate for G-TGase, due to the unfavorable amino acid sequences in their proximities. Chimeric proteins of IL-2 were therefore prepared by adding to the N-terminal sequence one of the following peptides known to be good G-TGase substrates: Pro-Lys-Pro-Gln-Gln-Phe-Met (called TG1), derived from Substance P [101] to give rTG1-IL-2, or Ala-Gln-Gln-Ile-Val-Met (called TG2), derived from fibronectin [102] to give rTG2-IL-2. The rTG1-IL-2 yielded one PEG chain conjugate (mono-PEGylated form), while

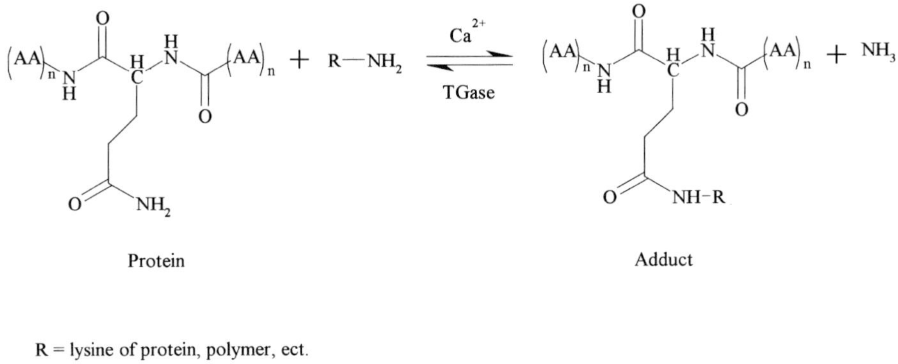

Scheme 6 Reaction catalyzed by TGase between a glutamine residue in a protein and an alkyl amine

the rTG2-IL-2 led to a mixture of mono-PEGylated forms and di-PEGylated forms. The enzymatic coupling of mPEG$_{3000}$-(CH$_2$)$_6$-NH$_2$ (PEG3) to rTG1-IL-2 was carried out in 0.1 M Tris – HCl buffer, pH 7.5 at 25 °C for 12 h in the presence of CaCl$_2$ 10 mM [103]. SDS-PAGE electrophoresis showed the prevalent formation of mono-PEGylated rTG1-IL-2. Moreover, it was demonstrated that intermolecular ε-(γ-glutamyl)lysine cross-linking does not take place for rhIL-2 and rTG1-IL-2 and that the native rhIL-2 is not a substrate for G-TGase. Incubation of rTG2-IL-2 (where double linkage is possible) with mPEG$_{10\,000}$-(CH$_2$)$_6$ – NH$_2$ (PEG10) under the same conditions, resulted in the formation of a mixture of mono- and di-PEGylated adducts, together with unreacted rTG2-IL-2 [103]. It was found that G-TGase can also work with high-molecular-weight PEGs, although at a lower yield, and can again form di-substituted derivatives at the two neighboring glutamines of rTG2-IL-2. TGase PEGylation gave derivatives with the same activity as the native protein, whereas the rTG2-IL-2 random modification with mPEG-NHS (10 kDa) yielded conjugates with an activity that is inversely related to the linked PEG mass (Table 5). On the other hand, it was found that the less specific M-TGase allowed for the incorporation of mPEG$_{12\,000}$-(CH$_2$)$_6$ – NH$_2$ (PEG12) directly into rhIL-2 [103]. The formation of a single conjugate, corresponding to the band at 35 kDa, was detected by SDS-PAGE electrophoresis. The apparent molecular weight of the conjugate was higher than expected, probably due to the bulkiness of PEG12-rhIL-2. A characterization carried out by LC–EMI/MS confirmed a mono-PEGylation at Gln-74, which also proved to be the unique site of incorporation when different alkylamines were employed [103]. Compared to other site-specific chemical PEGylation, such as thiol reactive PEG directly to a cysteine mutein or N-terminus modification at acidic pH by PEG-aldehyde, the enzyme coupling method produces fewer undesired products—namely, protein–protein dimers (due to cysteine oxidation) or εNH$_2$ lysine PEGylation (when N-terminus PEGylation is performed).

Table 5 Comparison of IL-2 conjugate activities between random PEGylation and direct-site PEGylation by TGase [100]

Proteins	% activity[a]
rhIL-2	100
PEG10-rhIL-2	74
(PEG10)$_2$-rhIL-2	36
rTG1-IL-2	72
PEG10-rTG1-IL-2	69
(PEG10)$_2$-rTG1-IL-2	72

[a] The amount of activity was expressed as a percentage of residual bioactivity as compared to rhIL-2

A two step enzymatic PEGylation, called GlycoPEGylation™, was developed by Neose Technologies. In this case, *E. coli*-expressed proteins (non-glycosylated) were first selectively glycosylated at specific serines and threonines in the protein amino acid sequence with a residue of *N*-acetylgalactosamine (GalNAc) by in vitro treatment with a recombinant O-GalNAc-transferase. The obtained glycosylated proteins were subsequently PEGylated using the O-GalNAc residue as the acceptor site for a sialic acid-PEG, a reaction selectively performed by a sialyltransferase [104]. The sialyltransferase transfers a cytidine monophosphate (CMP) derivative of PEGylated sialic acid (CMP-SA-PEG) at the level of O–GalNAc residue of glycosylated proteins. The great advantage of this technology is the possibility of obtaining PEGylated proteins that mimic their respective native proteins, since the PEG chains replace the sugar structure at the level of the native glycosylation site, therefore retaining the basic structure for receptor recognition plus and extended plasma half-life.

To underline the attention that this new system of conjugation is receiving, one may consider the increasing number of different enzymes studied for coupling peptides or polymers (beyond PEG) to protein, such as tyrosinase [105] and IgA protease [106].

Nonetheless, the applicability of these revolutionary enzymatic methods on an industrial scale has yet to be demonstrated and likely will encounter difficulties that will require time to overcome.

4.4
Protein PEGylation of Carboxylic Groups

PEGylation at carboxylic groups with an amino PEG is not an easy procedure, since undesired cross-linking may occur between a protein's activated COOHs and a protein's amino groups, yielding a number of unwanted side products.

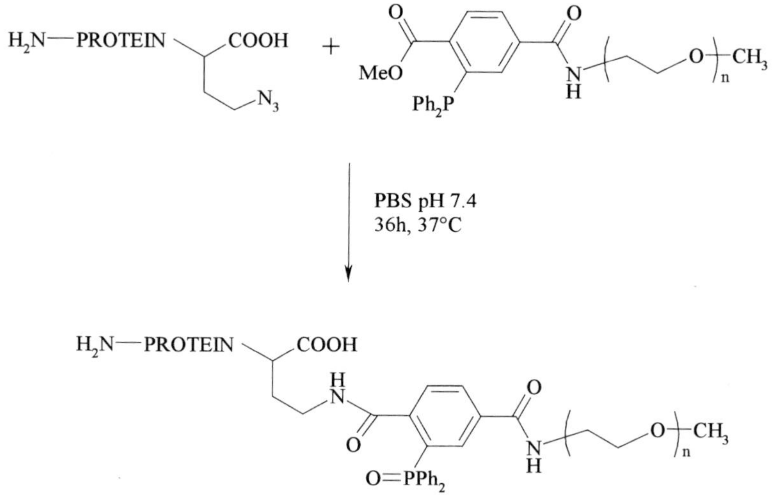

Scheme 7 Staudinger ligation leading to a C-terminal mono-PEGylated protein by reaction of a mutated protein containing a C-terminal azido-methionine with an engineered PEG derivative, methyl-PEG-triarylphosphine

One method proposed to overcome this problem involves the use of a PEG-hydrazide instead of the usual amino PEG. This allows the activation of the protein COOH groups by water-soluble carbodiimide at acidic pH in which all protein amino groups are protonated and rendered unsuitable for coupling, while the hydrazide group, which has a low pKa, can still react to form the polymer-protein adduct [107].

An innovative method for C-terminal site-specific PEGylation is based on Staudinger ligation [108]. The protocol, developed for a truncated thrombomodulin mutant [109], starts from the *E. coli* expression of a mutated protein containing a C-terminal azido-methionine, which reacts specifically with an engineered PEG derivative, methyl-PEG-triarylphosphine, leading to a C-terminal mono-PEGylated protein (Scheme 7). For this method, it is necessary to prepare a gene encoding for a protein with a C-terminal linker ending with methionine. Expression in *E. coli* is then induced only when the transformed bacteria are suspended in a medium where regular methionine is changed with its azido-functionalized analog. Unfortunately, this method is applicable only in the rare case of proteins devoid of methionine in the sequence; otherwise, they will stop the protein transduction because the azido analog does not permit the linking of further amino acids.

4.5
Beyond Protein PEGylation

PEGylation has not been limited to protein and non-protein drugs—recently, quite complex biological system such as cells and tissue have also been considered for the procedure.

PEGylation of red blood cells (RBCs) directed towards membrane proteins, carbohydrates, or lipid head groups, was the first example of this new research direction. It was devised mainly for transfusion purposes, but also for the preparation of stealth RBCs useful as drug delivery systems [110]. PEGylation leads to immunocamouflage of cells by covering antigenic sites and membrane surface charges, and by reducing receptor–ligand and cell–cell interactions. Scott and Chen obtained encouraging results in the PEGylation of white blood cells and later also with the more challenging pancreatic islets (containing the β-cells that produce insulin) [111]. The latter modification represents a milestone in the fields of cell and tissue PEGylation since the pancreatic cells have to retain, after PEG linking, not only the capacity to produce insulin but also the ability to fine-tune the release of the hormone following glucose response. The authors demonstrated the functionality of the Langerhans islets after PEG modification by transplanting the PEGylated cells into rats via portal vein injection for engraftment within liver vasculature.

5
Conclusion

Since the pioneering reports of Davis and Abuchowsky, described in the two best-known Journal of Biological Chemistry (JBC) papers [24, 112], an increasing number of research into PEG conjugation have been developed at academic institutions or pharmaceutical companies. The PEGylation concept, initially proposed and developed for protein modification [8, 9, 11], was later extended to peptide [10, 14], non-peptide drugs [113–116], and cells [111].

The initial years in the development of this technique have seen efforts to improve conjugation strategies by taking into consideration the most usual mild chemical strategies that could be compatible with the unstable structure of proteins. Various methods could be developed for conjugation to reactive groups in proteins. However, open problems still remain in this field—in particular, finding a satisfactory chemistry for selective binding at the level of the guanidine group of arginine, the indole group of tryptophan, or carboxylic groups. An area that requires further improvements is purification, which is an essential step in any PEGylation project. A typical example is the removal of unreacted PEG from the conjugate, which is particularly important for low-molecular-weight drugs, when small differences in weight and physical-chemical properties between conjugate and unreacted polymer make this separation difficult. Even more complicated is the separation of the many positional isomers that are usually present in the conjugation mixture. Moreover, methods to prevent the formation of several products at different extents of PEGylation, although accepted by the authorities (FDA), are also waiting for new and original solutions. Advancements in this direction may be represented by the recently described PEGylation based on enzymatic coupling as proposed by Sato et al. [100]. A request that is still expecting a satisfactory response is the availability of in vivo biodegradable PEG derivatives possessing one reactive point only. Last but not least, the problem of obtaining monodisperse, or at least very low polydisperse PEGs, especially in dealing with high-molecular-weight polymer species, still exists, despite the fact that LCC Engineering and Trading GmbH is now offering monodisperse PEGs on the market; unfortunately, these monodisperse PEGs apply only for low molecular weights—so far, below 600 Da.

Going to more general considerations, we are pleased to observe how PEG bioconjugation is now greatly expanding from proteins to non-peptide drugs and are being used to solve problems beyond immunogenicity and short residence time in blood, as it was in the early ages of PEGylation. In the area of non-peptide drugs, attention is now dedicated to the potential of the heterobifunctional PEGs, which allow for the combination of the advantages of polymer modification with the active targeting capacities of a second molecule linked to PEG. Thanks to the discovery of numerous new ligands [7] to target specific tissues or organs, or to entrap and release drug into cells, the field of non-peptide drug conjugation is receiving increasing interest. Of course, this area will need new ideas on the chemistry of binding to expand the number of the already existing heterobifunctional products offered on the market, mainly from Nektar Corporation.

An idea that will receive further development, although already described in many patents, is the combination of genetic engineering and PEGylation. This growing field is leading to conjugated muteins that have improved specificity or new properties with respect to native proteins or receptors. The patent filed by Genentech Inc [117] on GH receptor antagonist is a good example of developments in this arena.

One can of course ask why only PEG is chosen for protein modification instead of other polymers. Actually, very few examples of conjugation with other polymers have been reported so far, the most successful being the poly(styrene-co-maleic acid/anhydride) derivative of neocarcinostatin (SMANCS) developed by Maeda [118]. The main reason lies in the monofunctionality of mPEG that avoids a cross-linking reaction with the polyfunctionalized proteins. The usual natural or synthetic polymers present multiple points of attachment in the same molecule. This is the case of polysaccharide, for instance, but also for the extensively studied poly(N-(2-hydroxypropyl) methacrylamide) copolymers (HPMA) [119].

In conclusion, by reviewing hundreds of papers and patents, a selection of which has been presented here, the authors strongly believe that the field of PEGylation and conjugation through others polymers [12], is still a young research area in its early stages, although it is already considered a mature technology in many respects. Therefore, it should not be surprising to see new and unexpected applications coming from PEGylation in the near future.

Acknowledgements The authors thank the financial support of MURST 40%.

References

1. Davis FF, Abuchowski A, Van Es T, Palczuk NC, Chen R, Savoca K, Wieder K (1978) Enzyme Eng 4:169
2. Duncan R (2003) Nat Rev Drug Discov 2:347
3. Kopeček J, Kopeckova P, Minko T, Lu Z (2000) Eur Pharm J Biopharm 50:61
4. Harris JM, Chess RB (2003) Nat Rev Drug Discov 2:214
5. Veronese FM, Morpurgo M (1999) Farmaco 54:497
6. Harris JM, Martin NE, Modi M (2001) Clin Pharmacokinet 40:539
7. Davis FF, Kazo GM, Nucci ML, Abuchowski A (1990) In: Lee VHL (ed) Peptide and protein drug delivery. Dekker, New York, p 226
8. Nucci ML, Schorr R, Abuchowski A (1991) Adv Drug Deliv Rev 6:133
9. Russell-Jones GJ (1996) Adv Drug Deliv Rev 20:83
10. Okamoto CT (1998) Adv Drug Deliv Rev 29:215
11. Takakura Y, Mahoto Ri, Hashida M (1998) Adv Drug Deliv Rev 34:93
12. Allen TM (2002) Nat Rev Cancer 2:750
13. Bailon P, Berthold W (1998) Pharm Sci Technol Today 1:352
14. Zaplisky S, Harris JM (1997) In: Chemistry and biological applications of polyethylene glycol. American Chemical Society, San Francisco
15. Veronese FM (2001) Biomaterials 22:405
16. Levy Y, Hershfield MS, Fernandez-Mejia C et al. (1988) J Pediatr 113:312
17. Graham LM (2003) Adv Drug Deliv Rev 10:1293
18. Wang YS, Youngster S, Grace M, Bausch J, Bordens R, Wyss DF (2002) Adv Drug Deliv Rev 54:547
19. Bailon P, Palleroni A, Schaffer CA et al. (2001) Bioconjugate Chem 12:195
20. Reddy KR, Modi MW, Pedder S (2002) Adv Drug Deliv Rev 54:571
21. Trainer PJ, Drake WM, Katznelson L et al. (2000) N Engl J Med 342:1171
22. Kinstler O, Moulinex G, Treheit M et al. (2002) Adv Drug Deliv Rev 54:477
23. Eyetech Study Group (2002) Retina 22:143
24. Abuchowski A, Van Es T, Palczuk NC, Davis FF (1977) J Biol Chem 252:3578
25. Manjula BN, Tsai A, Upadhya R et al. (2003) Bioconj Chem 14:464
26. Working PK, Newman SS, Johnson J, Cornacoff JB (1997) Safety of poly(ethylene glycol) derivatives. In: Harris JM, Zalipsky S (eds) Poly(ethylene glycol) chemistry and biological applications. ACS, Washington, p 45
27. Petrak K, Goddard P (1989) Adv Drug Deliv Rev 3:191
28. Friman S, Egestad B, Sjovatt J, Svanvik J (1993) J Hepatol 17:48
29. Guiotto A, Canevari M, Pozzobon M, Moro S, Orsolini P, Veronese FM (2004) Bioorg Med Chem 12:5031
30. Enzon Inc (1999) US Patent 5 985 263
31. Applied Research System (1999) Wo Patent 9 955 376
32. Orsatti L, Veronese FM (1999) J Bioac Comp Polymer 14:429
33. Harris JM, Guo L, Fang ZH, Morpurgo M (1995) Seventh International Symposium On Recent Advances In Drug Delivery. Salt Lake City, USA
34. Monfardini C, Schiavon O, Caliceti P, Morpurgo M, Harris JM, Veronese FM (1995) Bioconjug Chem 6:62
35. Akiyama Y, Otsuka H, Nagasaki Y, Kato M, Kataoka K (2000) Bioconjugate Chem 11:947

36. Zhang S, Du J, Sun R et al. (2003) Reactive and functional polymer 56:17
37. Duncan R, Cable HC, Lloyd JB, Rejmanova P, Kopecek J (1984) Makromol Chem 184:1997
38. Rejmanova P, Pohl J, Baudys M, Kostka V, Kopecek J (1984) Makromol Chem 184:2009
39. Greenwald RB, Yang K, Zhao H, Conover CD, Lee S, Filpula D (2003) Bioconjug Chem 14:395
40. Greenwald RB, Choe YH, Conover CD, Shum K, Wu D, Royzen M (2000) J Med Chem 43:475
41. Choe YH, Conover CD, Wu D et al. (2002) J Control Release 79:55
42. Schiavon O, Pasut G, Moro S, Orsolini P, Guiotto A, Veronese FM (2004) European J Med Chem 39:123
43. Caliceti P, Schiavon O, Sartore L, Monfardini C, Veronese FM (1993) J Bioact Biocomp Polym 8:41
44. Esposito P, Barbero L, Caccia P et al. (2003) Adv Drug Deliv Rev 55:1279
45. Hooftman G, Herman S, Schacht E (1996) J Bioact Comp Polymer 11:135
46. Delgado C, Francis GE, Malik F, Fisher D, Parkes V (1997) Pharm Sci 3:59
47. Monkarsh SP, Ma Y, Aglione A et al. (1997) Anal Biochem 247:434
48. Veronese FM, Sacca B, Polverido, De Laureto P et al. (2001) Bioconjug Chem 1:62
49. Wylie DC, Voloch M, Lee S, Liu YH, Cannon-Carlson S, Cutler C, Pramanik B (2001) Pharm Res 18:1354
50. Uze G, Lutfalla G, Mongensen (1995) J Interferon Cytokine Res 5:3
51. Tyring SK (1995) Am J Obstet Gynecol 172:1350
52. Wang YS, Youngster S, Bausch J, Zhang R, Mcnemar C, Wyss DF (2000) Biochemistry 39:10634
53. Enzon Inc (1999) US Patent 5 985 263
54. Glue P, Fang JWS, Sabo R et al. (1999) Hepatology 30 (Suppl):A189
55. Pepinsky RB, Le Page D, Gill JA et al. (2001) J Pharmac Exp Ther 297:1059
56. Metcalf D (1986) Blood 67:257
57. Welte K, Gabrilove J, Bronchud MH, Platzer E, Morstyn G (1996) Blood 88:1907
58. Bowen S, Tare N, Inoue T, Yamasaki M, Okabe M, Horii I, Eliason JF (1999) Exp Hematol 27:425
59. Eliason JF, Greway A, Tare N et al. (2000) Stem Cells 18:40
60. Maxygen Holdings Ltd (2002) Wo Patent 0 236 626
61. Kinstler OB, Brems DN, Lauren SL, Paige AG, Hamburger JB, Treuheit MJ (1995) Pharm Res 13:996
62. Wong SS (ed) (1991) Reactive groups of proteins and their modifying agents. In: Chemistry of protein conjugation and cross-linking. CRC Press, Boca Raton, p 13
63. Lok S, Kaushansky K, Holly RD et al. (1994) Nature 369:565
64. Farese A, Hunt P, Macvitie TJ (1995) Blood 86:54
65. Kuter DJ, Beeler DLl, Rosemberg RD (1994) Proc Natl Acad Sci 369:11804
66. De-Boer RH, Roskos LK, Cheung E et al. (2000) Growth Factors 18:215
67. Guerra PI, Acklin C, Kosky AA, Davis JM, Treuheit MJ, Brems DN (1998) Pharm Res 15:1822
68. Chawla RK, Parks JS, Rudman D (1983) Annu Rev Med 34:519
69. Clark R, Olson K, Fuh G, Marian M et al. (1996) J Biol Chem 271:21969
70. Frohman LA, Jansson JO (1986) Endocr Rev 7:223
71. Lance VA, Murphy WA, Sueiras-Diaz J, Coy DH (1984) Biochem Biophys Res Commun 119:265
72. Piquet G, Gatti M, Barbero L, Traversa S, Caccia P, Esposito P (2002) J Chrom A944:141

73. D'Antonio M, Louveau I, Esposito P, Bartolino M, Canali S (2004) Growth Horm Igf Res 14:226
74. Ares Trading Sa (2002) Wo Patent 0 228 437
75. Goffin V, Bernichtein S, Carriere O, Bennett WF, Kopchick JJ, Kelly PA (1999) Endocrinology 140:3853
76. Parkinson C, Scarlett JA, Trainer PJ (2003) Adv Drug Deliv Rev 55:1303
77. King DJ (1998) Preparation, structure and function of monoclonal antibodies. In: Application and engineering of monoclonal antibodies. Taylor and Francis, London, p 1
78. King DJ, Adair JR (1999) Curr Opin Drug Discov Dev 2:110
79. Walsh G (2000) Nat Biotechnol 18:831
80. Chapman AP (2002) Adv Drug Deliv Rev 54:531
81. Koumenis IL, Shahrokh Z, Leong S, Hsei V, Deforge L, Zapata G (2000) Int J Pharm 198:83
82. Anderson WL, Tomasi TB (1988) J Immunol Methods 109:37
83. Maeda H, Wu J, Sawa T, Matsumura Y, Hori K (2000) J Control Release 65:271
84. Takashina K, Kitamura K, Yamaguchi T et al. (1991) J Cancer Res 82:1145
85. Pedley RB, Boden JA, Boden R, Dale R, Begent RH (1994) Br J Cancer 70:521
86. Wen X, Wu Qp, Lu Y, Fan Z, Charnsangavej C, Wallace S, Chow D, Li C (2001) Bioconjug Chem 12:545
87. Old LJ, Welt S (2003) US Patent 2003 031 671
88. Immunomedics Inc (1997) US Patent 5 670 132
89. Hinds KD, Kim SW (2002) Adv Drug Deliv Rev 54:505
90. Holtsberg FW, Ensor CM, Steiner MR, Bomalaski JS, Clark MA (2002) J Control Release 80:259
91. Amgen Inc (2003) US Patent 6 586 398
92. Amgen Inc (2002) US Patent 6 420 339
93. Hoffman La Roche (2000) US Patent 6 025 324
94. Applied Research System (1999) Wo Patent 9 955 377
95. Bolder Biotechnology Inc (1999) Wo Patent 9 903 887
96. Smith KA (1988) Science 240:1169
97. Goodson RJ, Katre NV (1990) Biotechnology 8:343
98. Chapman AP, Antoniw P, Spitali M, West S, Stephens S, King DJ (1999) Nat Biotechnol 17:780
99. Eistein Coll Med (1996) US Patent 5 585 484
100. Sato H (2002) Adv Drug Deliv Rev 54:487
101. Gorman JJ, Folk JE (1980) J Biol Chem 255:1175
102. Mcdonagh RP et al. (1981) Febs Lett 127:174
103. Sato H, Hayashi E, Yamada N, Yatagai M, Takahara Y (2001) Bioconjugate Chem 12:701
104. Neose Technologies Inc (2004) US Patent 2004 132 640
105. Chen T, Small DA, Wu LQ, Rubloff GW et al. (2003) Langmuir 19:9382
106. Lewinska M, Seitz C, Skerra A, Schmidtchen FP (2004) Bioconjug Chem 15:231
107. Zalipsky S, Menon-Rudolph Hydrazine S (1998) Derivatives 0f poly(ethylene glycol) and their conjugates. In: Harris JM (ed) Poly(ethylene glycol) chemistry: biotechnical and biomedical applications, 2nd edn. Plenum, New York, p 319
108. Saxon E, Bertozzi CR (2000) Science 287:2007
109. Cazalis CS, Haller CA, Sease-Cargo L, Chaikof EL (2004) Bioconjug Chem 15:1005
110. Murad KL, Mahany KL, Kuypers FA, Brugnara C, Eaton JW, Scott MD (1999) Blood 93:2121

111. Scott MD, Chen AM (2004) Transfusion clinique et biologique 11:40
112. Abuchowski A, Mccoy R, Palczuk NC, Van Es T, Davis FF (1977) J Biol Chem 252:3582
113. Greenwald RB (2001) J Control Release 74:159
114. Greenwald RB, Conover Cd, Choe Yh (2000) Crit Rev Ther Drug Carr Syst 17:101
115. Veronese FM, Schiavon O, Pasut G, Mendichi R, Andersson L, Tsirk A, Ford J, Wu G, Kneller S, Davies J, Duncan R (2005) Bioconjug Chem 16:775
116. Andersson L, Davies J, Duncan R, Ferruti P, Ford J, Kneller S, Mendichi R, Pasut G, Schiavon O, Tsirk A, Veronese FM, Vincenzi V, Wu G (2005) Biomacromol 6:914
117. Genentech Inc (2000) US Patent 6 057 292
118. Maeda H, Ueda M, Morinaga T, Matsumoto T (1985) J Med Chem 28:455
119. Duncan R, Kopeckova-Rejmanova P, Strohalm J, Hume I, Cable HC, Pohl J, Lloyd JB, Kopecek J (1987) Br J Cancer 55:165

Gene Delivery Using Polymer Therapeutics

Ernst Wagner (✉) · Julia Kloeckner

Pharmaceutical Biotechnology, Department of Pharmacy, Butenandtstrasse 5–13, 81377 Munich, Germany
ernst.wagner@cup.uni-muenchen.de, julia.kloeckner@cup.uni-muenchen.de

1	Introduction	137
2	Polymer Characteristics Required for Gene Delivery	137
3	Polyplexes	140
3.1	Commonly Used Cationic Polymers	140
3.1.1	Poly(L)lysine	140
3.1.2	Polyethylenimines	141
3.1.3	Polyamidoamine Dendrimers	142
3.1.4	Chitosan	143
3.1.5	Other Polymers	143
3.2	Optimization of Commonly Used/Established Cationic Polymers	146
3.2.1	Modified Polylysines	146
3.2.2	Modified Polyethylenimines	146
3.2.3	Modified Polyamidoamine Dendrimers	149
3.2.4	Novel Dendritic Polyamines	149
3.3	Biodegradable Polymers	151
3.3.1	Hydrolytic Degradation of Ester Bonds	152
3.3.2	Hydrolytic Degradation of Other Bonds	154
3.3.3	Reductive Cleavage of Disulfide Bonds	155
4	Incorporation of Delivery Functions	158
4.1	Targeting Domains	158
4.2	Shielding Domains	159
4.3	Transport Domains	161
5	Bioresponsive Polymers—Towards Artificial Viruses	164
6	Conclusions	167
	References	168

Abstract The use of polymers as synthetic non-viral carriers for introducing nucleic acids into cells appears very appealing. Polymers can be generated in large quantities in chemically defined, non-antigenic and non-immunogenic form. A plethora of different chemical structures and polymer sizes may be applied to tailor-made polymers with optimized characteristics for the extracellular delivery of nucleic acid to the target tissue and the subsequent intracellular delivery into the target cells. For the purpose of nucleic acid transfer, polymers have been applied for incorporating nucleic acids into nanoparticles or microspheres. Alternatively, cationic polymers are applied as carriers for complexing gene vectors into polyplexes. Polyplexes form spontaneously upon mixing negatively charged nucleic acid with the polycationic polymer due to electrostatic interaction. This process can be controlled to result in the formation of particles with defined virus-like sizes which efficiently transfect cell cultures and also have shown encouraging gene transfer potential in in vivo administration. With first-generation polymeric carriers, gene therapeutic effects have been demonstrated in animals, although modest efficiencies and significant toxicity restrict broader therapeutic application. Key issues for future optimization of polyplexes include improved specificity for the target tissue, enhanced intracellular uptake, and reduced toxicity and immunogenicity. Novel cationic polymers have to be made more biocompatible by reducing their potential for unspecific adverse interactions with the host, and by designing them in a biodegradable form. "Smart" polymers and polymer conjugates are being developed that in a dynamic manner present virus-like delivery functions in the appropriate phase of the gene delivery process.

Keywords DNA delivery · Non-viral vectors · Polycation · Polyplex · Targeted gene transfer

Abbreviations
ASGP	asialoglycoprotein
B-PEI	branched PEI
BSA	bovine serum albumine
CD	cyclodextrin
EGF	epidermal growth factor
HMW	high molecular weight
i.v.	intravenous
LLO	listeriolysin-O
LMW	low molecular weight
L-PEI	linear PEI
Mel	melittin
MW	molecular weight
N/P ratio	ratio of polymer nitrogens per DNA phosphate groups
n-PAE	network polyamino ester
NVP	N-vinyl-pyrrolidone
PAMAM	dendrimer, polyamidoamine dendrimer
pDMAEMA	poly(2-(dimethylamino) ethyl methacrylate)
pHPMA	poly-N-2-hydroxypropyl-methacrylamide
PEG	polyethylene glycol
PEI	polyethylenimine
PHLP	poly(D,L-lactide-co-4-hydroxyl-L-proline)
PLGA	poly(lactide-co-glycolide)
pLL	poly(L)lysine
PPE-PA	poly(2-aminoethyl propylene phosphate)
PPI	polypropylenimine
PVA	poly(vinyl alcohol)
RPC	reducible polycation
TNF	tumor necrosis factor

1
Introduction

Non-viral vectors are receiving increasing attention as gene delivery systems because of several favorable characteristics. They offer enhanced biosafety and biocompatibility, a high flexibility regarding the size of the delivered nucleic acid, and can be synthesized at low cost in large quantities and with high consistency of production. Such delivery systems, when able to deliver therapeutic genes to the target tissues in vivo with high efficiency and specificity, would be very attractive for a broad variety of biomedical applications.

For the purpose of nucleic acid transfer, amongst other synthetic carrier systems, polymers [1] have been applied for the incorporation of DNA into nanoparticles [2–5] or polyplexes [6–10]. The current chapter focuses on the development and characteristics of polyplexes which form spontaneously upon mixing negatively charged nucleic acid with cationic polymer due to electrostatic interaction.

Cationic polymers have been developed and used for non-viral gene delivery systems for nearly four decades [11]. Polymers can be generated in chemically defined, non-toxic, non-antigenic and non-immunogenic form. In theory, a plethora of different chemical structures and polymer sizes may be applied to tailor-made polymers with optimized characteristics for delivery of nucleic acid, such as highly specific transport to the target tissue, and efficient subsequent delivery into the nuclei of target cells. The chapter outlines key requirements to be met by the applied polymers and polyplexes, it reviews the results obtained so far and ongoing strategies to further improve polymer-based gene transfer.

2
Polymer Characteristics Required for Gene Delivery

For effective polyplex-mediated gene delivery, the cationic polymer carrier has to fulfill a series of drug delivery functions in the extracellular and intracellular transport of the DNA vector (see Fig. 1b). The polymer has to compact DNA into particles of virus-like dimensions that can migrate through the blood circulation into the target tissue, it has to protect the DNA from degradation and against undesired interactions with the biological environment, to facilitate target cell binding and internalization [12–16], ideally in a target cell-specific manner [17, 18], endosomal escape [19], trafficking the cytoplasmic environment [20] and localizing into the nucleus [21, 22] as well as vector unpacking [23]. In addition, the polymer should be non-toxic, non-immunogenic, and biodegradable. In reality, no polymer is able to carry out all the extracellular and intracellular delivery functions; therefore additional functional elements have to be included into the polyplex formulation (see Fig. 1b).

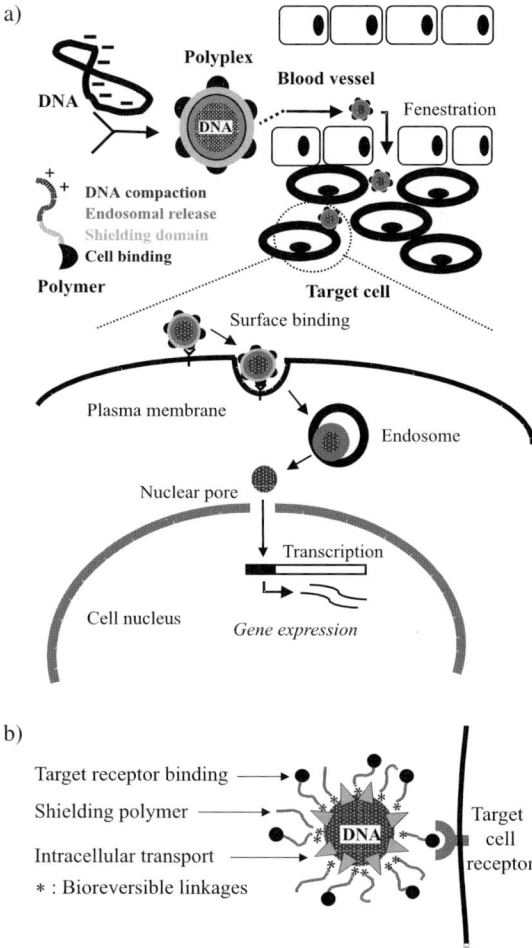

Fig. 1 Polymer characteristics for in vivo application. **a** The polymer should be non-toxic and non-immunogenic, form polyplexes which are stable and inert in blood, protect DNA against extracellular and intracellular degradation, allow transport to the target tissue, specific internalization into the target cell, and intracellular trafficking into the cell nucleus. **b** To introduce specificity and enhance efficiency, delivery functions may be conjugated to the polymer which facilitate the different steps

The basic function of the cationic carrier is binding of DNA and its compaction into particles. Factors influencing DNA binding affinity which are inherent to the chemical structure of the polymer ("intrinsic properties") are: (*i*) the number of charge groups per single polymer; (*ii*) the type of charge groups (e.g. primary, secondary, quaternary amino groups, amidine groups); (*iii*) the spacing of charge groups within the polymer; (*iv*) the degree of branching in the polymer backbone; and (*v*) hydrophobicity of the cationic carrier.

In addition, the "external/extrinsic" conditions of the microenvironment, such as: (*vi*) the ionic strength of the polyplex solution; (*vii*) the concentration and (*viii*) positive/negative charge ratios of polymer and DNA; and (*ix*) the technical process of polyplex formation (i.e. kinetically vs. thermodynamically controlled process) strongly influence the polyplex formation. The latter "extrinsic" factors provide some flexibility in selecting the most appropriate conditions for polyplex formation. However, it has to be kept in mind that finally upon administration the physiological environment will dictate the stability and fate of polyplexes; therefore the proper intrinsic polymer characteristics in DNA binding and polyplex formation are the dominating issues.

In regard to the required number of positive polymer charges, systematic studies demonstrated that a minimum length of six to eight cationic amino acids (lysine, arginine) is required to compact DNA into polyplex structures active in gene delivery in vitro [24]. The number is presumably higher when in vivo application is considered. DNA binding can be driven by application of higher polymer to DNA charge ratios. This can be nicely monitored by agarose gel retardation assay or ethidium bromide exclusion assay [25]. Depending on size and affinity of the polycation this may also result in increased net positive polyplex charge which promotes cellular uptake and transfection efficiency. However, this also results in toxicity, due to destabilization and loss of integrity of cellular membranes, and the presence of excess free polycation.

The influence of charge group type [26] and spacing [27] was evaluated by Davis and colleagues in a carbohydrate-containing polycation series. They demonstrated that both the distance between the carbohydrate unit and the charge groups in the backbone, and also the types of amino group (quaternary amines vs. amidine group) are the primary factors that influence the carrier's transfection efficiency in vitro. On BHK-21 cells quaternary amines showed similar toxicity but lower gene expression than the amidine analogues. The presence of chloroquine enhanced the transfection activity of quaternary amines but not for the amidine.

The degree of polymer branching has a significant effect on the flexibility of these macromolecules and hence their ability to complex and transport DNA. For example, in evaluating histidine/lysine copolymers (HK) in combination with a liposomal carrier, Chen et al. [28] showed that the degree of branching was a major factor determining the transfection efficiency. In transformed cell lines, branched HK polymers were significantly more effective than the linear HK polymer, however, linear HK enhanced gene expression in primary cell lines more effectively. The differences in the optimal carrier (linear vs. branched) were not due to initial cellular uptake or size of the complexes. There was a strong association between the optimal type of HK polymer and the pH of endocytic vesicles. In cell cultures with the linear HK polymer showing the best effects, endocytic vesicles were strongly acidic with a pH below 5. Conversely, in the cell lines in which the branched polymers were optimal transfection agents, the pH of endocytic vesicles was above 6.

3
Polyplexes

Over the last four decades many different polycations have been employed in polyplexes, including natural DNA binding proteins such as histones, the synthetic amino acid polymers such as polylysine, polyornithine, and other cationic polymers such as polyamidoamine dendrimers, polyethylenimines, chitosan, polyphosphoramidates, or poly (dimethylaminoethyl) methacrylates. The use of these and other polymers has been previously reviewed in [7–9, 29, 30]. The characteristics of polymers which have been previously most commonly used are discussed below (Sect. 3.1), followed by the review of strategies to optimize these polymers (Sect. 3.2) or novel biodegradable polymers (Sect. 3.3).

3.1
Commonly Used Cationic Polymers

3.1.1
Poly(L)lysine

Poly(L)lysine (pLL) was one of the first polycations used for polyplex formation. At physiological pH the amino groups of pLL are positively charged and interact ionically with the negatively charged phosphate groups of the DNA, condensing DNA into toroid-like structures of around 100 nm [31]. pLL/DNA polyplexes have a highly positive zeta potential, interact electrostatically with a negatively charged cell surface, and are taken up by absorptive endocytosis. The efficiency of uptake can be enhanced by covalently coupling pLL with ligands that can specifically target cells and promote receptor-mediated uptake [29].

For targeted delivery to the hepatocyte-specific asialoglycoprotein (ASGP) receptor, DNA/asialoorosomucoid-pLL complexes were administered which resulted in gene expression in rat liver [32]. Expression was transient but could be significantly prolonged by partial hepatectomy performed 30 minutes after injection of the polyplex [33]. Subsequent work successfully applied the system for hepatocyte-specific gene transfer of the human albumin gene to nagase analbuminemic rats [34] and the LDL receptor gene in a rabbit animal model of familial hypercholesterolemia [35], which resulted in a temporary amelioration of the disease phenotypes. Applying a similar ASGP receptor targeting concept [36], expression of coagulation factor IX in the liver was observed after application of DNA/galactose-pLL complexes into the caudal vena cava of rats for up to 140 days. The condensation of DNA into polyplexes of small particle size of around 20 nm was considered an important factor for prolonged expression.

Davis and colleagues developed several strategies for pLL-mediated gene transfer to the lung. For targeting to the polymeric immunoglobulin receptor, they generated small-sized anti-pIg Fab-pLL polyplexes. Systemic delivery of anti-pIg polyplexes in rats resulted in reporter gene expression in cells of the airway epithelium and submucosal glands [37]. In an alternative approach, using a synthetic peptide ligand derived from alpha 1-antitrypsin for targeting the serpin-enzyme complex receptor, polyplexes were optimized regarding the ligand to pLL ratio and molecular weight of pLL [38]. Long-chain pLL (54 kDa) generated smaller (25 nm) polyplexes than short-chain pLL (10 kDa; generating 40 nm polyplexes) and the smaller particles gave significantly higher and longer duration of gene expression in vivo.

Despite these encouraging results, in general the gene expression using pLL polyplexes was rather low. It was found that intracellular uptake of polyplexes was quite effective, but subsequent escape from intracellular vesicles into the cytoplasm presents a major bottleneck. For example, transferrin-pLL DNA complexes were efficiently internalized into transferrin receptor-expressing K562 cells, but entrapment in intracellular vesicles prevented gene expression [39]. The addition of the lysosomotropic agent chloroquine to the medium was shown to increase transfection efficiency by more than 1000-fold, probably by interfering with lysosomal degradation and enhancing the release of the DNA into the cytoplasm.

The inclusion of other endosomolytic agents has been shown to dramatically enhance gene transfer; addition of replication-defective viruses (adenovirus or rhinovirus) with endosomal membrane-destabilizing properties augments the levels of gene transfer more than 1000-fold [19, 40]. To broaden the applicability of these findings, endosomolytic agents were directly attached or incorporated into the DNA polyplexes (see Sect. 4.3). Recently an alternative novel technique was developed to improve endosomal release: the light-induced photochemical rupture of endocytic vesicles [41, 42]. This method termed photochemical internalization (PCI) has already shown encouraging results in improving transfection efficiency of polylysine polyplexes [41, 43–45].

3.1.2
Polyethylenimines

Currently, polyethylenimines (PEI) are most frequently used because of an excellent transfection efficiency in vitro and significant transfection in vivo [46, 47]. In addition, PEIs are inexpensive reagents available in large quantities and in various forms. Cationic PEI polyplexes enter cells through adhesion to negatively charged transmembrane heparanproteoglycans [15]. In contrast to polylysine, PEI can promote its escape from intracellular vesicles. This capability is thought to be based on its "proton sponge" effect, which contributes to the intrinsic ability of PEI to facilitate endosomal release [48]. PEI can change its degree of protonation depending on the surrounding pH, for instance only one out of six nitrogen atoms is protonated at neutral pH in the absence of DNA; polyplex-bound PEI is also only partially protonated (approx. every second nitrogen). For complexation of DNA, usually a N/P (PEI nitrogen/DNA phosphate) charge ratio of 6 or higher is applied, generating polyplexes with considerable buffering capacity at lower endosomal pH. Upon intracellular delivery of the DNA particle, the acidification process within the endosome triggers protonation of complex-bound and when present also free PEI, inducing chloride ion influx, osmotic swelling and destabilization of the vesicle [48].

The problematic characteristics of PEI are a pronounced toxicity both in cell culture and in vivo and a whole variety of undesired unspecific interaction with the biological environment. Despite an efficient systemic gene transfer to the lung using positively charged polyplexes [47, 49], the in vivo application remains restricted because of severe acute toxicity including death [50, 51]. Therefore, topical gene transfer of PEI polyplexes is being explored as an alternative delivery route, such as instillation or nebulization for delivery to the lung [52, 53].

The type of PEI polymer (molecular weight, branched, or linear form) and the polyplex formulation can largely influence both the toxicity and the gene transfer efficiency [54–56]. For example, purification of PEI polyplexes by size exclusion chromatography [56] recently demonstrated that purified polyplexes, devoid of free PEI, have strongly reduced toxicity both in vitro and in vivo, but also significantly reduced transfection activity; free PEI apparently contributes to overcome the bottleneck of endosomal release.

Cell culture experiments [54] showed that large-sized PEI/DNA polyplexes (aggregates of several hundred nm) are more effective but more toxic. The endosomal escape is thought not to be the major bottleneck for such large particles. Small PEI/DNA particles, around 50–100 nm, however, have lower efficacy, resulting from an inefficient endosomal release. Photochemical internalization (PCI) is one method to enhance efficiency of the small PEI polyplexes [43, 57].

3.1.3
Polyamidoamine Dendrimers

The first type of polycationic molecules described with high transfection potential in the absence of endosomolytic agents was the polyamidoamine (PAMAM) polymer ("Starburst") dendrimer synthesized with terminal amino groups [58–61]. The dendrimer shows high transfection potential in the absence of any additional membrane-active agent. Due to the low pKa of its terminal and internal amines this polymer can act as an endosomal buffering agent [48] preventing this way DNA degradation within the endolysosomes. For optimum efficiency, rather high dendrimer to DNA charge ratios (e.g. amine: phosphate ratio of 6 : 1) have to be used. The positive charge of the polyplex is considered to be required both for endosomal release and electrostatic cell surface binding. Manunta et al. [62] investigated whether the internalization of dendrimer/DNA polyplexes takes place randomly on the cell surface or at preferential sites such as membrane rafts. Using an endothelial cell line model, polyplexes were found to co-localize with the membrane ganglioside GM1. Binding and especially internalization of dendrimer-DNA complexes was strongly reduced by cholesterol depletion before transfection. The data suggest that membrane cholesterol and raft integrity are physiologically relevant for the cellular uptake of dendrimer-DNA complexes.

The positive surface charge of the polyplex can present problems, associated with cellular toxicity and further aspects relevant for in vivo applications, similarly as described for PEI (see above).

3.1.4
Chitosan

Chitosan, (poly-D-glucosamin, generated by deacetylation of chitin) is a nontoxic biodegradable polycationic polymer with low immunogenicity. Chitosan can effectively bind DNA and protect it from nuclease degradation [63].

Various modified chitosans have been applied for gene transfer. Lactosylated chitosan was reported to mediate transfection efficiency to HeLa cells which is comparable to PEI [64]. Galactosylated chitosan grafted with PEG forms stable polyplexes which have low transfection efficiency [65]. Amines of oligomeric chitosan were quaternized by methylation. Use of trimethylated chitosan oligomers [66] resulted in efficient transfection of COS-1 and CaCO-2 cells.

The receptor ligand transferrin has been incorporated into chitosan/DNA polyplexes which enhanced gene transfer up to four-fold compared to unmodified chitosan [67]. In a similar fashion, incorporation of C-terminal domain of adenovirus fiber knob protein enhanced transfection up to 130-fold in HeLa cells. Further modifications include the incorporation of hydrophobic moieties to generate dodecylated chitosan, deoxycholic acid modified chitosan. Urocanic acid-modified chitosan [68] was reported to mediate efficient gene delivery; it was hypothesized that the imidazole ring plays a crucial role for enhancing the release of internalized polyplexes from endosomal vesicles.

3.1.5
Other Polymers

Polymethacrylates

The group of Hennink et al. [69–72] synthesized and evaluated poly(2-(dimethylamino)ethyl methacrylate) pDMAEMA for gene transfer. The polymer forms positively charged DNA polyplexes which can be used successfully for in vitro transfection of different cell lines, including COS-7 and OVCAR-3 cells. HMW forms of the polymer (> 300 kDa) were able to condense DNA effectively into particles of 150–200 nm, whereas LMW pDMAEMA forms large complexes (size 0.5–1.0 micrometer). Like other cationic polymers, pDMAEMA is cytotoxic. Therefore, copolymers of DMAEMA with methyl methacrylate (MMA), ethoxytriethylene glycol methacrylate (triEGMA), or N-vinyl-pyrrolidone (NVP) were also evaluated. A copolymer with 20 mol % of MMA showed a reduced transfection efficiency but increased cytotoxicity. A copolymer with triEGMA (48 mol %) showed both a reduced transfection efficiency and a reduced cytotoxicity, whereas a copolymer with NVP (54 mol %) showed an increased transfection efficiency but a decreased cytotoxicity as compared to pDMAEMA.

Cell trafficking experiments showed that—like for many other polymeric transfection reagents—the endosomal escape is a limiting bottle-neck. The use of cationic polymers with a pKa around or slightly below physiological pH has been considered as a possible way to enhance endosomal escape ("proton sponge" see above, Sect. 3.1.2). Henninck and colleagues [73] synthesized a new polymer pDAMA, with two tertiary amine groups in each monomeric unit, poly(2-methyl-acrylic acid 2-[(2-(dimethylamino)-ethyl)-methyl-amino]-ethyl ester). One pKa of the monomer is approximately 9, providing cationic charge at physiological pH, and thus DNA binding properties; the other pKa is approximately 5 and provides endosomal buffering capacity. pDAMA has a low toxicity but also very low transfection activity. The addition of a membrane-disruptive peptide considerably increased the transfection efficiency. This indicates that the pDAMA polyplexes alone are not able to mediate escape from the endosomes via the proton sponge mechanism, which implies that the proton sponge hypothesis is not always applicable for polymers with buffering capacity at low pH.

The same lab also evaluated the fate of pDMAEMA polyplexes upon systemic application in vivo [72]. Their data indicate that aggregate formation of positively charged polyplexes with blood components followed by trapping of the polyplex aggregates in the lung capillaries is probably responsible for a preferential lung uptake and transfection. Strategies explored by the researcher to overcome these undesired characteristics are reviewed in Sect. 4.2.

Cyclodextrins

Davis and colleagues generated linear cationic beta-cyclodextrin-based polymers (betaCDPs) as gene transfer carriers [74]. The initial betaCDPs were synthesized by the condensation of a diamino-cyclodextrin monomer A with a diimidate monomer B. The inclusion of a cyclodextrin moiety in the monomer A units reduces the IC50s of the polymer by up to 3 orders of magnitude. The spacing between the cationic amidine groups is also important. Different polymers with 4, 5, 6, 7, 8, and 10 methylene units (betaCDP4, 5, 6, 7, 8, and 10) in the monomer B molecule showed up to a 20-fold difference in transfection efficiency between polymers. Optimum transfection was achieved with the betaCDP6 polymer.

Recently, more data [26, 27] were reported evaluating the influence of the cationic charge group type and carbohydrate spacing on the transfection efficiency (see also Sect. 2, above). The same group [75] describes a cyclodextrin-based gene delivery system that is surface-modified to display poly(ethylene glycol) (PEG) for increasing stability in biological fluids and transferrin for targeting to cancer cells that express the transferrin receptor. A transferrin-PEG-adamantane conjugate self-assembles with the nanoparticles by adamantane (host) and particle surface cyclodextrin (guest) inclusion complex formation. The particles remain stable in physiological salt concentrations and transfect K562 cells with increased efficiency over untargeted particles. The increase in transfection is eliminated when transfections are conducted in the presence of excess free transferrin.

Another recently published approach applies oligocationic cyclodextrins (CDs) that were modified with pyridylamino, alkylimidazole, methoxyethylamino, or primary amine groups at 6-positions of the glucose units [76]. The oligocationic CDs neutralized DNA to form stable nanoparticulate polyplexes. The transfection efficiency of these CDs was dependent on the substituents present, with the most efficient having either an amino, pyridylamino, or butylimidazole group at the 6-positions. One of the most effective vectors, heptakis-pyridylamino CD, produced a 4000-fold increase in transfection level over DNA alone. Levels were improved 10-fold by use of the endosomolytic agent chloroquine. The transfection efficiency equals that of DOTAP. Uptake studies indicate that the polycationic CDs efficiently promote cellular uptake of DNA, dependent on proteoglycan-mediated binding to cells. Also with these formulations intracellular trafficking presumably is the rate-limiting step in the transfection process.

CDs were also applied to optimize the established polycationic gene transfer carriers PAM dendrimer and PEI, see [77–79] and the relevant sections in our chapter.

Pluronic Polymers

Kabanov and colleagues [80] applied pluronic polymers for gene transfer. These are PEO-PPO-PEO block copolymers that are not cationic in nature, consisting of various ratios of ethylene oxide and propylene oxide. Such a non-ionic carrier composed of two amphiphilic block copolymers, pluronics L61 and F127, increased intramuscular expression of plasmid DNA about 10-fold [81]. Additionally, efficient delivery of DNA was found with low concentrations of the copolymer, which further reduces toxicity. The same lab also synthesized a conjugate of 2 kDa LMW PEI with the Pluronic 123 [82]. In combination with free P123 and DNA the conjugate forms 110 nm small and stable complexes that after i.v. injection into mice exhibit high gene expression in the liver. In a similar approach, 2 kDa PEI grafted with Pluronic P85 was applied as a carrier for antisense oligonucleotide delivery [83]. Subcutaneous treatment of athymic nude mice bearing subcutaneous human HT29 colon adenocarcinoma xenografts, combined with a single dose of irradiation, caused a significant inhibition of tumor growth compared with mismatch- and naked antisense-pretreated control groups.

Polylactide-Hydroxyproline

Li and Huang [84] investigated another copolymer, poly(D,L-lactide-*co*-4-hydroxyl-L-proline) (PHLP). This biocompatible copolymer displays lower toxicity as compared to PLL or PEI. When applied in DNA microspheres, it was found to have increased gene expression over longer periods of time.

3.2
Optimization of Commonly Used/Established Cationic Polymers

Optimization of cationic polymers is based on the following motivation: to improve the gene transfer efficiency, to make carriers less toxic and more biocompatible. Strategies include the optimization and modification of commonly used polymers such as polylysine or polyethylenimine (see below, this section); and the generation of novel, biodegradable polymers (see Sect. 3.3).

3.2.1
Modified Polylysines

Poly(L)lysine can efficiently bind DNA and, especially in the presence of receptor-binding ligands, deliver polyplexes into intracellular vesicles, but it does not mediate endosomal escape. To overcome this limitation, Midoux and colleagues [85] substituted polylysine with histidyl residues, with the rationale that the imidazole group of histidine would serve as a protonable endosomal escape moiety. Consistent with the "proton sponge" hypothesis, histidinylated polylysine mediated much higher in vitro transfection than polylysine. Similarly, Putnam et al. [86] conjugated imidazole groups to the epsilon-amines of polylysine. In this case the transfection efficiency was improved with increasing imidazole content in the polymers. The polymer with the highest imidazole content mediated gene expression levels similar to those mediated by PEI, but with less cytotoxicity [86].

3.2.2
Modified Polyethylenimines

PEI has been found to be a very effective transfection reagent as outlined above, but its use is limited by its toxicity (LD50 of L-PEI in Balb/C mice is 4 mg/kg). Therefore, either optimization of the PEI polyplex structure and/or the development of new biocompatible conjugates is required. A series of modifications have been made, not only to reduce toxicity and enhance efficiency, but also to increase the understanding on the gene transfer mechanism [87]. The major directions in modifications are reviewed below.

PEG-Grafted PEI

Systemic administration of polyplexes of unmodified 22 kDa L-PEI results in very high gene expression in the lung [47, 49, 55]. However, in vivo studies have shown that there is a narrow window between efficiency and severe toxicity [50]. Other PEI molecules may even do worse; for example, in a report by Kircheis et al. [51] severe acute toxicity was observed after systemic application, with 50% of the mice dying with clinical signs of acute lung embolism. PEI-based polyplexes are positively charged to be effective. They aggregate rapidly in physiological salt conditions and due to their charge they unspecifically interact with erythrocytes and other blood components [88].

To reduce the unspecific binding capacity of PEI, conjugates of PEI with PEG were synthesized [89–93]. For example, PEG chains of various sizes (5, 20, or 40 kDa) were conjugated with 22 kDa L-PEI or 25 kDa B-PEI [91, 92]. These "shielding" conjugates were mixed at various ratios with unmodified L-PEI or B-PEI as the DNA condensing agent, EGF-PEG- or Tf-PEG-modified B-PEI (as the targeting agent, see below), followed by mixing with plasmid DNA. The mixing ratios of the PEI conjugates (usually 10% targeting and 20–30% PEG conjugates) and PEI (usually 60–70%) strongly influence the biophysical characteristics of the formed DNA particles. PEGylation strongly reduces the zeta potential to $<+5$ mV, improves solubility of the DNA complexes, stabilizes against aggregation of DNA particles, and strongly reduces the toxicity in vitro and in systemic delivery (see Sect. 4.2). Among other in vivo applications, PEG-modified PEI polyplexes have been successfully evaluated for topical delivery to the nasal epithelium [90] and for local delivery to the muscle after application in a lipiodolized emulsion [93].

Dextran-Grafted PEI

As an alternative to a hydrophilic PEG shield, dextrans (with MW 10 kDa or 1.5 kDa) were used for grafting on either 25 kDa L-PEI or 25 kDa B-PEI at various molar degrees [94]. Only the branched B-PEI-dextran conjugate was found to stabilize DNA polyplexes in the presence of BSA, as evaluated by measurement of size, zeta potential, and lack of DNA release (DNA release analyzed by TO-PRO-1 intercalation into DNA).

Cyclodextrin-Grafted PEI

β-Cyclodextrin has been known to increase solubility of various formulations and also can form inclusion complexes with adamantane derivatives. To exploit this property, L-PEI and B-PEI were grafted with β-cyclodextrin [79]. The resulting conjugates, β-CD-L-PEI and β-CD-B-PEI, respectively, were combined with adamantane-conjugated PEG (AD-PEG) before DNA polyplex formation. The β-CD/AD-PEG inclusion complexes which anchor the PEG polymers to the DNA/PEI particle surface, are formed, providing particle stabilization and gene transfer activity in the absence of toxicity. Independent from the presence of PEG, both toxicity and also transfection efficiency of CD-PEI polyplexes was reduced by increasing the CD-grafting. Possible reasons might include reduced endosomal release or reduced endosomal buffering capacity of the modified PEIs, outlining requirements that have to be dealt with in the further optimization process.

Other PEI Modifications

The molecular weight and backbone structure (linear or branched) of PEI, and the presence of residual protective groups from the synthesis (e.g. propionamide residues deriving from poly-2-ethyl-2-oxazoline) or related modifications have been found to strongly influence the transfection characteristics [87, 95, 96]. Chemical modifications have been introduced onto the nitrogen atoms, such as acetylation or dodecylation, introduction of hydroxyethyl groups, or quarternization by methylation [97–100]. In some cases these modifications led to PEI derivatives with markedly enhanced performance and/or reduced toxicity. For example, dodecylation of 2 kDa PEI yields a non-toxic polycation whose transfection efficiency in the presence of serum is 400-fold enhanced. Acetylation of 25 kDa B-PEI [100] decreased the endosomal buffering capacity, defined as the moles of protons absorbed per mole of nitrogen on titration from pH 7.5 to 4.5. Acetylation, surprisingly, resulted in increased gene delivery effectiveness in MDA-MB-231 and C2C12 cell lines up to 21-fold compared to unmodified PEI, both in the presence and absence of serum. The mechanism is not yet understood, but the enhancement may be caused by more effective polyplex unpackaging for DNA release, altered endocytic trafficking, and/or increased lipophilicity of acetylated PEI-DNA complexes.

PEI-cholesterol lipopolymers were generated by grafting B-PEIs (1.8 kDa and 10 kDa) with one cholesterol molecule per PEI molecule [101]. Coupling of cholesteryl chloroformate was performed via secondary amines of PEI, to allow interaction of PEI's primary amines with the DNA. The polymer conjugate had the preferred high buffering capacity within the range pH 5–7, and mediated efficient DNA condensation. Efficient gene transfer, however, was decreased in the presence of serum. Furgeson and colleagues [102] described the modification of L-PEI with cholesterol. Another hydrophobic modification was reported by Sochanik et al. [103] who described gene transfer in vitro as well as in vivo using cetylated 0.6 kDa LMW PEI (28% of amine groups substituted with cetyl moieties) incorporated into cholesterol liposomes. Highest in vivo luciferase expression was observed in the lungs.

3.2.3
Modified Polyamidoamine Dendrimers

Cyclodextrin-Grafted Dendrimer

To improve the dendrimer transfection efficiency, Uekama and colleagues [77] synthesized the starburst PAMAM dendrimer conjugates with alpha-, beta-, and gamma-cyclodextrins (CD conjugates) in an equimolar ratio of 1 : 1. CD conjugates showed a potent luciferase gene expression, especially the alpha-CD conjugate which provided the greatest transfection activity (approximately 100-times higher than of the unmodified dendrimer, and superior to that of lipofectin. An optimized formulation was also evaluated for in vivo gene transfer [78] demonstrating favorable characteristics in comparison to the native dendrimer.

Arginine-Grafted Dendrimer

The primary amines located on the surface of PAMAM dendrimer were conjugated with L-arginine [104]. Polyplexes composed of L-arginine-grafted PAMAM/DNA showed increased gene delivery potency compared to native PAMAM dendrimer and L-lysine-grafted PAMAM, as evaluated for 293, HepG2, and Neuro 2A cells and in primary rat vascular smooth muscle cells.

3.2.4
Novel Dendritic Polyamines

Dendritic Polylysine

Dendritic poly(L-lysine) of the 6th generation (KG6) [105] and the analogues KGR6 (with the terminal amino acids replaced by arginine) and KGH6 (terminal lysines replaced by histidine) [106] were generated and applied for polyplex formation. DNA binding studies revealed that both KGR6 and KG6 bind strongly, whereas KGH6 shows lower binding ability. With regard to transfection efficiency, KGR6 mediated higher transfection efficiency than KG6, and KGH6 showed no transfection efficiency unless polyplexes were formed under acidic conditions.

In comparison to PEI polyplexes or standard polylysine polyplexes, dendritic poly(L-lysine) KG6 polyplexes appear to have more favorable biophysical characteristics for systemic administration; a low zeta potential (+ 3 mV) of particles might be responsible for prolonged circulation in the blood flow. Passive tumor targeting was observed starting at 1 h after administration. About 1% of the applied DNA dose was observed in the tumor [107]. In vivo gene expression data have not yet been reported for this system.

Dendritic Polypropylenimine

Different generations of polypropylenimine (PPI) dendrimers (generations 1 to 5: DAB 4, DAB 8, DAB 16, DAB 32, and DAB 64) have been evaluated as gene delivery systems [108]. Cell cytotoxicity was largely generation dependent and followed the trend DAB 64 > DAB 32 > DAB 16 > DAB 4 > DAB 8, whereas transfection efficacy followed the trend DAB 8, DAB 16 > DAB 4 > DAB 32, DAB 64. The generation 2 polypropylenimine dendrimer DAB 8 combines a sufficient level of DNA binding with a low level of cell cytoxicity to give it optimum in vitro gene transfer activity.

DAB 16 polyplexes showed very encouraging results upon systemic administration into tumor-bearing mice. Apparently, the low molecular weight of 1.7 kDa of PPI prevented undesired unspecific effects of the polymer, such as blood aggregation or gene expression in the lung. Intravenous injection of marker gene formulations delivered gene expression to solid tumors, and therapeutic polyplexes encoding TNF-alpha upon systemic delivery resulted in complete regression of A431 tumors [109].

Hyperbranched PEI Dendrons

Novel hyperbranched dendron polymers were synthesized using a 10 kDa LMW B-PEI core [110]. Using successive attachment of ethyleneimine moieties to the PEI core, the relative ratio of linear-to-branched structures was lowered from 1.17 to 0.70. The more extensive branching of PEI enables the condensation of plasmid DNA into nanostructures of small size (70–100 nm), stability at least for 3 weeks at 4 °C, and a very low cytotoxicity in vitro. Under optimized conditions, the transfection activity at a N/P ratio of 6 was approximately six times higher than that of the commercially available PEI transfection reagent. Bioluminescent imaging of in vivo gene expression using a luciferase reporter gene showed gene expression in the liver and in the lymph nodes of mice.

Other Dendritic Polyamines

Similarly to that described above, Krämer et al. [111] investigated the influence of molecular weights and degrees of branching on the transfection efficiency and the cell toxicity of cationic polymers. Functionalization of 6 kDa to 25 kDa B-PEI by a two-step procedure generated fully branched pseudodendrimers: analogues of polypropylenimine (PPI) and polyamidoamine (PAMAM) dendrimers. The cytotoxicity of the dendrimers generally rises with increasing core size. It was optimum 6 kDa for sensitive and 21 kDa for robust cell lines. The DNA transfection efficiencies observed for these polymers also depended on the cell line (NIH/3T3 and COS-7 cells). The highest efficiencies were observed for polymers whose PEI cores had molecular weights in the range of 6–25 kDa. A maximum transfection efficiency was observed at 60% branching for the PPI analogues, and at a 25 kDa B-PEI core.

3.3
Biodegradable Polymers

Polycationic carriers are toxic because of their avidity to unspecifically bind not only to negatively charged DNA but also to the many other slightly negatively charged biological materials including phospholipid bilayer cell membranes, cell membrane-integrated and circulatory proteins, or for example erythrocytes. These toxic properties increase with the number of positive charges and hydrophobicity of the polycation. Neutralization of the positive polymer charges by polyplex formation reduces toxicity, both on the cellular level and in an organism, however, does not eliminate the problem. In addition to acute toxicity, the long-term fate of the polymeric carrier has also to be considered in the host.

Therefore, biocompatible and biodegradable polymers, which can be degraded by the host would be advantageous. Partial or complete degradation of the polycation results in the reduction of positive charges per molecule which is also expected to reduce unspecific interactions and toxicity. In fact, many low molecular weight (LMW) polymers have been found to possess strongly reduced toxicity compared to their high-molecular weight (HMW) counterparts, see for example [111–113], however, do not provide sufficient stability of the polyplexes for in vivo administration. For these reasons two basic strategies are being explored in the development of novel biocompatible cationic polymers: (*i*) the synthesis of new polymers which have biodegradable bonds within their repetitive monomeric unit; or alternatively, (*ii*) generation of biodegradable polymer conjugates where LMW polymers with low toxicity are crosslinked into larger polycationic carriers by conjugation with biodegradable linkers. As outlined in detail below, the biodegradable bonds may be hydrolytically labile groups such as esters or phosphoesters, acetals, or hydrazones; alternatively they can be disulfide bonds which are cleavable within reducing cellular compartments. Other biochemical cleavage points may also be designed, such as target sites for enzymatic degradation by specific proteases or esterases.

3.3.1
Hydrolytic Degradation of Ester Bonds

Poly (Amino Acid) Ester Analogues

Lim et al. [114] synthesized a biodegradable ester analog of polylysine, poly[alpha-(4-aminobutyl)-L-glycolic acid] (PAGA). While the polymer displayed no cytotoxicity, only modest transfection activity was observed. This may be due to a too fast hydrolysis and, similar to polylysine, the lack of efficient endosomal escape functionality. Another biodegradable cationic polyester, poly (4-hydroxy-L-proline ester) showed similar characteristics [115].

Succinate-linked PEG—pLL Copolymer

Multi-block copolymers were synthesized based on LMW pLL (1.5–6 kDa) copolymerized with difunctional 1.5 kDa PEG succinate ester [113]. Synthesized copolymers showed almost negligible cytotoxicity and transfection efficiency was comparable to the PLL homopolymer with a MW of 26 kDa. Biodegradation of the succinate ester linkages under physiological conditions revealed that the molecular weight of copolymers decreased to 20% of the initial MW within 72 h. Transfection efficiencies of copolymers were not affected by the presence of serum, while that of PLL homopolymer decreased to the level of naked DNA in the presence of serum.

Poly (β-Amino Ester)

In a combinatorial approach, Anderson et al. [116] synthesized 2350 structurally unique, degradable cationic polymers. The synthesis was based on Michael addition of various primary amines or secondary di-amines with a variety of di-acrylates in DMSO. Without further purification, a large-scale transfection screen was performed, identifying candidates with interesting gene transfer properties.

The same laboratory [117] applied this conjugation chemistry for the defined synthesis of two poly(β-amino esters), generated by the addition of 1-aminobutanol to 1,4-butanediol diacrylate or 1,6-hexandioldiacrylate. By variation of the amine: di-acrylate ratio, the nature of the end groups and the molecular weight (3.4–18 kDa) was modulated. Maximum transfection activity on COS-7 cells was obtained using amino-terminated polymers of about 13 kDa MW and applying high charge ratio (polymer/DNA ratio of 30–100 w/w).

Network Poly (β-Amino Ester)

Recently, Lim et al. [118] described a new biodegradable polymer: a branched network of amino esters (n-PAE) which has transfection efficiencies similar to PEI 25 kDa, however with minimal cytotoxicity. The high transfection efficiency was attributed to the proton sponge effect in endosomes similar to that described for PEI. The network structure of the polymer is based on polycondensation of TRIS molecules N-disubstituted with methyl acrylate, and terminal amino groups were attached to the polyester condensate in the form of 6-amino hexanoic acid esters. This structure provides multiple ternary and primary amines for DNA binding and endosomal buffering. The network structure is also important to tune polyester degradation to intermediate stability, while linear amino-modified polyester appears to show too fast hydrolysis rates.

Amine-Modified Graft Polyesters

Kissel and colleagues recently reported the development of biodegradable comb-branched polymers [119] consisting of amine-modified PVA. The PVA backbone was grafted with PLGA side chains as spacers for various amine modifications: 3-diethylamino-1-propylamine, 2-diethylamino-1-ethylamine, and 3-diethylamino-1-propylamine. The PLGA side chains as spacer mediate fast polymer degradation, i.e. release of the cationic charge groups. Polymer degradation was monitored by NMR at 37 °C in PBS. Compared to linear PLGA the amine-modified polyester degraded more rapidly as shown by the reduction of the length of the side-chain.

Dimethylaminoethanol Methacryloylamino Carbonic Ester (pHPMA-DMAE)

A new biodegradable polymer pHPMA-DMAE, poly [carbonic acid 2-dimethylamino-ethyl ester 1-methyl-2-(2-methacryloylamino)-ethyl ester] was synthesized [120] which forms small polyplexes (110 nm) with a positive zeta potential, similar to that found for the analogue pDMAEMA polyplexes previously reported by the same research group. Degradation of the polyplexes at 37 °C and pH 7.4 or pH 5 was monitored. Interestingly, intact DNA was released from the polyplexes after 48 h at pH 7.4, whereas all DNA remained bound to the polymer at pH 5.0. Polyplexes were able to transfect cells with minimal cytotoxicity, if the endosomal membrane-disrupting influenza-derived synthetic peptide INF-7 [121] was added to the polyplex formulation.

Biodegradable PEI Derivatives Synthesized from LMW PEI Core Units

Biodegradable PEI-PEG-copolymers [122] were synthesized derived from LMW PEI (0.6, 1.2, or 1.8 kDa) and hydrophilic bifunctional 2 kDa PEG-di (succinimidyl succinate). Copolymers containing succinate ester linkages were degradable under physiological conditions and non-toxic. In vitro transfection efficiency of the synthesized copolymer increased up to three times higher than that of the native LMW PEI.

Poly(ethylenimine-co-L-lactamide-co-succinamide) was synthesized as a copolymer of 1.2 kDa LMW-PEI and 1 kDa oligo(L-lactic acid-co-succinic acid) [123]. The resulting copolymer P(EI-co-LSA) (8 kDa) forms small polyplexes (150 nm) with a positive surface charge of + 18 mV. Similar to LMW PEI, the copolymer exhibited a low toxicity profile; P(EI-co-LSA) degrades via base-catalyzed hydrolytic cleavage which is higher at pH 7 than at pH 5.

New degradable PEI derivatives with low toxicity and efficient gene transfer activity were synthesized by Pack and colleagues [124]. LMW PEI (0.8 kDa) was crosslinked with molar equivalents of diacrylates of either 1,3-butanediol or 1,6-hexanediol, yielding 14–30 kDa branched PEIs, Poly-1 or Poly-2, respectively. The crosslinking reaction is based on Michael addition to the amino groups of PEI which does alkylate but not eliminate protonable amino groups. Degradation halflifes at 37 °C were 4 h and 30 h for Poly 1 and Poly 2, respectively, and the hydrolysis rate was not different at pH 5 or 7. The transfection activity of these essentially non-toxic materials was much higher than LMW PEI and up to 16-fold higher than for 25 kDa B-PEI.

Polyphosphoesters and Polyphosphoramidates

Polyphosphoester [125–127] and polyphosphoramidates [30, 128] present another class of biodegradable gene carriers. Biodegradability is based on hydrolytic cleavage of phosphoester bonds at physiological pH.

Polyphosphoester poly(2-aminoethyl propylene phosphate), (PPE-PA), was synthesized and evaluated for polyplex formation. Stability of DNA polyplexes increased with molecular weight of the polymer. Enhanced in vivo gene expression in mouse muscle was observed based on sustained release from the PPE-PA carrier [127].

Polyphosphoramidates bearing a spermidine side chain (PPA-SP), with approx. 100 positive charges per polymer chain [128] were able to condense plasmid DNA efficiently and showed lower cytotoxicity than PLL and PEI in cell culture. Gene expression mediated by PPA-SP was greatly enhanced when chloroquine was used as an endo/lysosomal protective and escape agent. Optimized PPA-SP/DNA complexes yielded gene expression levels close to PEI/DNA complexes.

3.3.2
Hydrolytic Degradation of Other Bonds

Hydrolysis of Ortho Esters, Acetals, or Hydrazones

Intracellular pH gradients can be exploited in the cellular delivery of macromolecules [129]. Polyorthoesters (POEs) are relatively stable under physiological pH, but rapidly hydrolyze at the endosomal/lysosomal pH of 5. This characteristic has been exploited for enhanced delivery of DNA vaccines by incorporating DNA into poly(ortho ester) microspheres [130], at pH 5, 100% of the DNA was released within 24 h.

Another acid-degradable linker has been incorporated into polymers by Murthy et al. [131]. Terpolymers have been generated that consist of a hydrophobic, membrane-disruptive poly(propylacrylic acid) backbone [132], onto which hydrophilic PEG chains have been grafted via an acid-degradable acetal linkage. Providing a pH-responsive membrane-disruptive component, this polymer was shown to direct the uptake and endosomal release of oligonucleotides in cultured hepatocytes.

Walker et al. [133] reported the use of pH-sensitive hydrazone linkers for linking PEG to a polycationic carrier. Polyplexes containing such a bioreversible PEG-polycation conjugate upon exposure to endosomal pH undergo deshielding by PEG removal within less than 1 h at 37 °C (see also below, Sect. 5).

3.3.3
Reductive Cleavage of Disulfide Bonds

Several strategies for intracellular delivery of nucleic acids are based on reductive cleavage of disulfide bonds [134–136]. Disulfide bonds are known to be stable in the blood, but are cleaved inside the cell; this is believed to be due to the fact that the concentration of glutathione, the most abundant reducing agent in most cells including mammals, is in a millimolar range inside the cell but in a micromolar range in blood plasma. It is also known that protein disulfide reduction occurs late in the endocytic pathway including lysosomes [137]. Therefore, once inside the cell, disulfide linkages should be reduced, leading to the cleavage of disulfide-containing HMW polymers into LMW fragments which are lower in toxicity.

LMW Disulfide Cross Linking Peptides

Rice and colleagues [138, 139] generated LMW disulfide cross-linking DNA carrier peptides consisting of lysine and two terminal cysteine residues that, when bound to DNA, polymerize through disulfide bond formation. This results in small, stable polyplexes that mediate efficient in vitro gene transfer. Substitution of histidine for lysine residues resulted in an optimal peptide of Cys-His-(Lys)$_6$-His-Cys that also provided buffering capacity to enhance gene expression. Extending this strategy, targeting glycopeptides and shielding PEG molecules were incorporated [140]. Recently, the same group also examined in vivo gene delivery of sulfhydryl cross-linked PEG-peptide/targeting glycopeptide DNA polyplexes, following i.v. dosing in mice [141]. Optimal targeting to hepatocytes was achieved by condensing DNA with a mixture of 10 mol % targeting glycopeptide, 5 mol % PEG-peptide, and 85 mol % backbone peptide. Optimized gene delivery formulations transiently expressed secreted alkaline phosphatase in mouse serum for 12 days [141].

Disulfide Cross-Linking Oligolysines

Oupicky et al. [142] demonstrated that crosslinking short polylysine in polyplexes with a bioreducible crosslinking agent increased the stability of polyplexes. Masking the surface with PEG, a 10-fold increased in vivo plasma circulation was observed after intravenous administration to mice. Subsequently Oupicky et al. [143] generated linear polylysine-type reducible polycations (RPCs) of 187 kDa and 45 kDa by oxidation of the terminal cysteinyl thiol groups of $Cys(Lys)_{10}Cys$. These polycations were used for the formation of polyplexes which were coated and surface-cross-linked using multivalent reactive copolymers of pHPMA for steric stabilization. Cell culture transfections of lipopolyplexes containing combinations of RPC with DOTAP liposomes resulted in up to 190-fold higher gene expression levels as compared to control experiments using the non-reducible PLL/DOTAP lipopolyplexes [144]. This very clearly indicated the relevance of intracellular bioreductive release for gene transfer efficiency.

Trubetskoy and colleagues [145] introduced disulfide crosslinks into polylysines by treatment with dimethyl-3-3′-dithiobispropionimidate (DTBP) after complex formation. This step enhances the stability of polylysine polyplexes ("caged DNA polyplexes"), as demonstrated by lack of aggregation at high ionic strength. The bonds stabilize polyplexes in an extracellular medium and are bioreducible in intracellular compartments. It is thought that they should not hinder subsequent intracellular release to facilitate access of DNA to the nuclear transcription apparatus.

On the basis of analogous thoughts, block catiomer polyplexes were developed by Miyata et al. [146] by controlling both the cationic charge and disulfide cross-linking densities of the backbone polycations. A PEG-PLL block copolymer was thiolated using either of two thiolation reagents, N-succinimidyl 3-(2-pyridyldithio)propionate (SPDP) or 2-iminothiolane (Traut's reagent), to investigate the effects of both the charge and disulfide cross-linking densities. The introduction of thiol groups by SPDP proceeded through the formation of amide linkages to concomitantly decrease the cationic charge density of the PLL segment, whereas Traut's reagent promoted the thiolation with the introduction of cationic imino groups which keep the charge density constant. These thiolated PEG-PLLs were complexed with pDNA to form approximately 100 nm disulfide cross-linked polyplexes. Both thiolation methods were similarly effective in introducing disulfide cross-links to prevent the polyplex from the dissociation through a counter polyanion exchange. On the other hand, in the reductive condition mimicking the intracellular environment, an efficient release of pDNA was only achieved for the polyplex thiolated with SPDP, which has the lower cationic charge density. This polyplex also revealed approximately 50-times higher transfection efficiency than the polyplex thiolated with iminothiolane. Obviously, the balance between the densities of the cationic charge and disulfide cross-linking in the thiolated polyplex played a crucial role in the delivery and controlled release of DNA to achieve high transfection efficiency.

Poly(Lys-(AEDTP))

A new cationic polymer was synthesized by modification of PLL amino groups with 3-(2-aminoethyldithio)propionyl residues [147]. This modification replaces the positive amino charge groups of the lysine residues by more distal aminoethyl charge groups. Poly(lys-(AEDTP)) formed polyplexes of 100 nm and had a zeta potential of + 17 mV. Upon incubation with reducing agents the polymer loses the charge groups, triggering dissociation of the polyplex. Transfection with poly(lys-(AEDTP)) is more efficient than with the non-disulfide-linked PLL, which is consistent with the hypothesis that intracellular release of DNA from polyplexes can be beneficial for the gene transfer process.

Disulfide-Crosslinked LMW PEI

Bioreversible crosslinking by disulfide bridges has also been investigated by Gosselin et al. [148] using LMW PEI. PEI of 0.8 kDa was crosslinked with the bifunctional crosslinkers dithiobis(succinimidylpropionate) (DSP) or dimethyl-3-3'-dithiobispropionimidate (DTBP). Using DSP and DTBP, the primary amino groups of PEI which are crosslinked are converted into amide or amidine groups, respectively. The resulting PEI conjugates showed improved transfection activity compared to the starting LMW PEI, but were less effective than 25 kDa B-PEI. Importantly, bioreversible conjugates have significantly reduced cytotoxicity as compared to 25 kDa B-PEI.

In summary, the literature describes the crosslinking of 0.8 kDa LMW PEI with three different bioreversible linkers: the two disulfide forms as described above [148], and diacrylate-crosslinked PEI as mentioned in Sect. 3.3.1 and described in [124]. Apart from the biodegradable linkage (disulfide vs. ester) the conjugates differ in the type of amino group formed at the linkage sites: the PEI amino groups are converted either into (uncharged) amide bonds (SP linkage), into positively charged amidines (IP linkage), or into (protonable) beta-aminoester groups (HD linkage). Different characteristics may also rise from differential modification of either primary or secondary PEI amines. Comparing the data in the literature, it was unclear which of the forms mediates the best transfection activity. Therefore, we compared the two disulfide forms of crosslinked LMW PEI with diacrylate-crosslinked LMW PEI (Kloeckner and Wagner, our unpublished results). Figure 2 shows that the highest efficiency was observed using the hexanediol diacrylate crosslinked PEI (HD linkage) especially when polyplexes with only a moderate excess of positively charged polymer were applied. Using a large excess of polycation increased the efficiency of the two other forms (IP and SP linkage); it seems

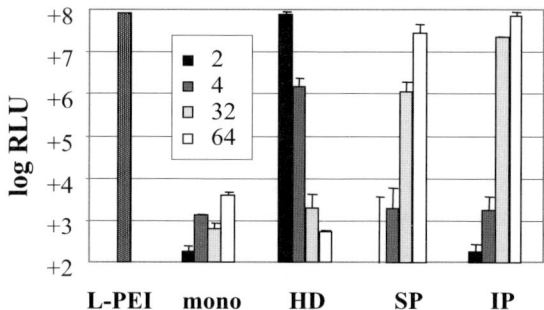

Fig. 2 Gene transfer using biodegradable PEI derivatives. Polyplex carriers were synthesized from 0.8 kDa LMW PEI core units using hexanediol-diacrylate as the linker (generating HD-PEI) in an analogous fashion to that described in [117], or the bifunctional disulfide-containing crosslinkers DSP or DTBP (generating SP-PEI or IP-PEI) as described in [148]. Luciferase gene expression of polyplexes formed at different polycation/DNA (w/w) ratios containing 200 ng pCMVL plasmid DNA and using B16F10 cells (5000 cells/well) is shown. L-PEI of 22 kDa at optimum N/P charge ratio of 6 was used as the gold standard; "mono", starting monomer 0.8 kDa LMW PEI; HD, HD-PEI conjugate; SP, SP-PEI conjugate; IP, IP-PEI conjugate

that in the latter case the presence of large amounts of free polycation (not bound to the DNA polyplex) is required for the transfection process.

4
Incorporation of Delivery Functions

Without special modifications, polyplex formulations cannot distinguish between target and non-target tissue. Cell-binding ligands have to be incorporated into polyplexes which recognize and bind target cell-specific receptors (see Sect. 4.1). In addition, modifications have to be made to shield polyplex domains (see Sect. 4.2) which otherwise induce undesired binding to blood components or non-target cells (see also Fig. 1b).

4.1
Targeting Domains

Numerous cell-targeting ligands have been incorporated into polyplexes after chemical conjugation to cationic polymers. Such cell-binding ligands can be small molecules and vitamins [149, 150], carbohydrates [151, 152], peptides [38], or proteins [18, 32] including growth factors [153–155] or antibodies [156–159]. Detailed reviews on evaluated conjugates are available in [160, 161]. Successful targeting in cell culture has been described demonstrating up to 1000-fold enhanced gene expression in target cells in comparison to transfection controls like with ligand-free complexes.

In vitro target specificity is required but not sufficient for in vivo targeting, because interactions with other (non-target) cells, plasma components, or the extracellular matrix inhibit targeting. In contrast to the in vitro studies, only a few studies report successful in vivo targeting [34, 36, 162] using ligand-containing polyplexes. In many cases the systemic administration of polyplexes through the tail vein in mice resulted in high acute toxicity and the highest gene expression in the lung tissue; see [51]. Non-specific interactions of polyplexes and aggregation with blood components are the major reasons for these effects. Plank et al. [163] observed that positively charged polyplexes activate the alternative pathway of the complement system which is part of the innate immune system. Formation of erythrocyte agglomerates [88] is another undesired side effect. Such non-specific interactions of polyplexes with blood components followed by trapping of polyplex aggregates in the lung capillaries are responsible for the preferential lung transfection and toxicity [50, 72].

4.2
Shielding Domains

To make ligand-mediated targeting effective and more specific, polyplex domains with unspecific binding activity have to be masked to obtain the desired targeting specificity. Hydrophilic polymers like polyethylene glycol (PEG) or hydroxypropyl methacrylate (pHPMA) have been attached to the DNA polyplex surface. For the attachment of hydrophilic polymer to the polycation several strategies have been developed. Covalent coupling of the hydrophilic polymer was performed either before polyplex formation ("prePEGylation") [91, 154, 164] (see also Sect. 3.2.2) or after the polyplex formation ("postPEGylation") [88, 164, 165], optionally also incorporating cell targeting ligands. Another approach described a prePEGylation strategy to display PEG and the targeting ligand transferrin by a non-covalent adamantane (host)/cyclodextrin (guest) inclusion complex formation [75]. A further approach utilized the transferrin which as serum protein is well adapted to the requirements within the blood circulation for both surface shielding and targeting [162].

Shielding provides not only increased solubility and improved stability for freeze-thawing [92], but also reduced toxicity and extended circulation time in blood. Applying such shielding strategies, systemic in vivo targeting of tumors was demonstrated in mice (see Fig. 3). Intravenous injection of Tf-coated or Tf/PEG-coated polyplexes resulted in gene transfer into distant subcutaneous Neuro2A neuroblastoma tumors of syngeneic A/J mice [88, 91, 162]. In an analogous manner, EGF-PEG-coated polyplexes were successfully applied for systemic targeting of human hepatocellular carcinoma xenografts in SCID mice [154]. In these models, luciferase marker gene expression levels in tumor tissues were 10- to 100-fold higher than in other organ tissues. Re-

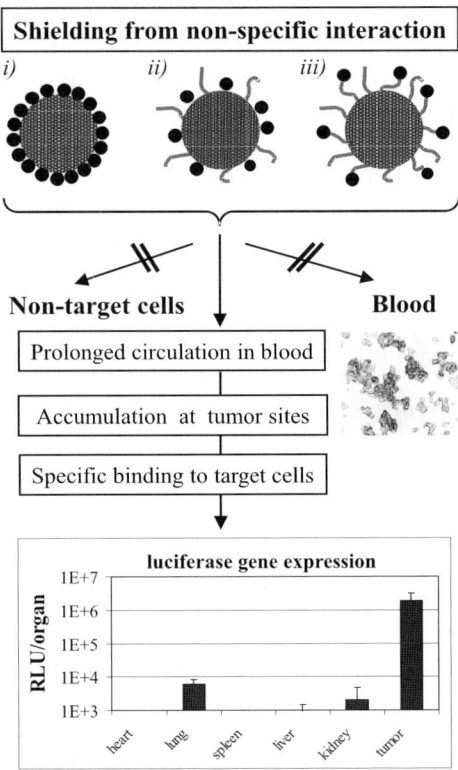

Fig. 3 Shielding of Tf-PEI/DNA complexes from non-specific interactions for systemic in vivo application. Polycation/DNA complexes with cell-binding ligands (e.g. transferrin) can specifically interact with target cells expressing the according receptor. Non-specific interactions with blood components and non-target cells can be blocked by shielding the surface charge of the transfection complexes applying (*i*) high ligand density (e.g. high transferrin), (*ii*) Post-PEGylation of ligand-PEI/DNA complexes, or (*iii*) Pre-PEGylation, i.e. use of PEG-PEI conjugates together with ligand-PEG-PEI conjugates, with the ligand at the top of the PEG coat. Shielding enables a prolonged circulation time of polyplexes in the blood, leading to accumulation at sites of higher vascular leakiness, e.g. tumor sites, and followed by uptake and gene expression in tumor cells [162]

peated systemic application of Tf-coated polyplexes encoding tumor necrosis factor alpha (TNF-alpha) into tumor-bearing mice induced tumor necrosis and inhibition of tumor growth in four murine tumor models of different tissue origin [91, 166]. As gene expression of TNF-alpha was localized within the tumor, no systemic TNF-related toxicities were observed.

Similar to the PEI polyplexes mentioned above, Verbaan and colleagues [164] generated pDMAEMA polyplexes with the surface charge effectively shielded by two PEGylation methods: prePEGylation, i.e. the use of pDMAEMA-graft-PEG polymers and postPEGylation of preformed complexes. The shielding effect was the highest for the postPEGylation method with 20 kDa PEG, yielding polyplexes which showed little interaction with blood components (i.e. albumin and erythrocytes) and showed substantially prolonged circulation time in mice after i.v. administration. This translated into tumor accumulation of about 3.5% of the injected dose per gram tumor tissue in the subcutaneous Neuro2A tumor model and to about 4.2% of the injected dose per gram tumor tissue in a subcutaneous C26 tumor model. However, no significant gene expression levels were observed, indicating that PEGylation may interfere with subsequent intracellular delivery steps.

One possible reason for the lack of gene expression might be the absence of a targeting ligand such as transferrin. However, surface charge shielding also markedly reduces gene transfer in the case of PEI polyplexes. This can only partly—but not completely—be restored by incorporating a targeting ligand such as transferrin. Bioreversible incorporation of a shield into polyplexes is considered as one encouraging approach to recover efficiency; recent strategies in this direction will be reviewed in Sect. 5.

4.3
Transport Domains

Once the cell has been targeted, the vector particle has to be internalized by the cellular uptake machinery via endocytotic or phagocytotic uptake processes. This step can be taken rather easily by polyplexes, as appropriate receptor-binding ligands and/or cationic charges may enhance intracellular uptake of particles into endosomal vesicles. Several intracellular barriers then have to be overcome for successful transgene expression. Endosomal release was found to be a major bottleneck for many non-viral vectors [151, 167]. The vector particle needs to survive and escape from the endosomal vesicular compartment, traffick the cytoplasmic environment, target the nucleus, enter the nucleus, and expose the carried nucleic acid to the cellular transcription machinery.

Particle size, surface charge, and ligands all influence cellular binding, uptake and intracellular trafficking of polyplexes [12–16, 62]. Rejman et al. [168] report the size-dependent internalization of particles into non-phagocytic tumor cells via different pathways: irrespective of surface charge or ligand, small particles (< 200 nm) internalize via clathrin-mediated endocytosis, whereas large particles (500 nm) internalize via caveolae-mediated endocytosis. Positively charged particles have been reported to bind and internalize after binding to negatively charged transmembrane heparan proteoglycans [12]. In a recent study using PEI polyplexes, Kopatz et al. [15] described a model for non-viral entry of cationic polyplexes through adhesion to specific transmembrane heparan proteoglycans called syndecans followed by clathrin-independent internalization of vesicles by the actin cytoskeleton. Colocalization of PEI polyplexes with the actin cytoskeleton was demonstrated. It remains unclear whether the latter pathway contributes to efficient gene delivery or rather represents a "dead end". Goncalves et al. [14] found that both clathrin-dependent pathways of polyplexes and clathrin-independent macropinocytosis are supposed to be the productive pathways mediating the delivery of genes. The macropinocytosis pathway, however, was found to impair transfection efficiency. Ruponen and colleagues [16] describe that the uptake of DNA complexes depends on the carrier, cell type and the amounts of the polyanionic glycosaminoglycans (GAGs) heparan sulfate, chondroitin sulfate and hyaluronan on the cell-surface. However, all cell-surface GAGs inhibit the transgene expression and probably direct complexes into intracellular compartments that do not support gene transfer.

After uptake of DNA complexes, escape from intracellular vesicles into the cytoplasm represents a major bottleneck. Entrapment in lysosomal or phagocytic vesicles is thought to be associated with degradation of the complexes in these compartments. The fate of the delivered DNA strongly depends on the selected polycationic carrier. Capture in endo/lysosomes seems to be a more serious hurdle for pLL polyplexes than for dendrimer or PEI polyplexes. The following section will discuss approaches to overcome this barrier.

Endosomal Escape

Biological events like entry of viruses and toxins into the cells, the action of antibacterial peptides, defense toxins, or the performance of complement, defensins, and perforins of the vertebrate immune system are examples demonstrating how efficiently nature can modulate membrane barriers. A series of natural membrane-destabilizing agents have been characterized at the molecular level [169, 170]. In several cases the membrane-active principle was found to be located in defined, small amphipathic peptide domains. Such domains were incorporated into DNA polyplexes to enhance their escape from the endo/lysosomal vesicular compartment into the cytoplasm [40, 167, 171–173]. Apart from virus particles, proteins such as adenovirus penton protein [174], bacterial cytolysines [175], or the transmembrane domain of diphtheria toxin [176] have been used as endosomolytic agents. In addition, synthetic peptides with sequences derived from viral sequences have been tested. For instance the N-terminus of influenza virus hemagglutinin HA-2, the N-terminus of rhinovirus HRV2 VP-1 protein, and other natural or artificial sequences such as the amphipathic peptides GALA, EGLA, KALA, JTS1, melittin or gramicidin S have been investigated [58, 121, 151, 177, 178]. PLL-mediated gene transfer can be improved up to more than 1000-fold by acidic membrane-active compounds. Some other polycations like dendrimers or PEI have inherent endosomal escape activity and are only moderately enhanced by endosomolytic peptides.

For example, Plank et al. [151] generated pLL polyplexes containing as the targeting ligand a synthetic tetra-antennary carbohydrate ligand for hepatic ASGP receptor binding, and as a membrane-active peptide a synthetic acidic peptide analogue derived from the N-terminal HA-2 subunit of the influenza virus hemagglutinin to induce pH-specific endosomal release. Both components were covalently attached to pLL as the DNA-binding element. Application of these particles to cultured hepatocytes resulted in efficient, ligand-specific gene expression which was highest with inclusion of the endosome-destabilizing peptide.

Hashida and colleagues [152] extended this concept for in vivo application. They developed a polyplex system consisting of polyornithine, which was modified first with galactose to serve as the ASGP receptor ligand, then with a fusogenic peptide derived from the influenza virus HA2 domain. Upon intravenous injection in mice, a large amount of transgene product was detected in the liver, and the hepatocytes contributed to more than 95% of total tissue gene expression.

Studies by Mechtler and Wagner [121] showed that for pLL-mediated gene transfer the use of acidic influenza peptide versions with specificity for endosomal acidic pH generated the best results. Such enhancing effects were also observed with other polymers, as described for example in [73, 120]. For lipoplex-mediated gene transfer, less acidic variants gave better results [179].

PEI polyplexes were only moderately enhanced by influenza virus-derived peptides; the most pronounced effects were obtained when small-sized PEI polyplexes were used [54]. Ogris et al. covalently attached the cationic peptide melittin via its N-terminus to PEI [178, 180]. While free uncoupled melittin displayed significant toxicity, the N-terminal linked melittin-PEI conjugate displayed low toxicity and enhanced reporter gene expression within a broad range of cell lines and types tested; even slowly dividing or non-dividing primary cells were susceptible to transfection. Coupling melittin via the N-terminus might be important for the positive function. The natural form of melittin is assumed to form pores after inserting the N-terminus into the lipid membrane; obviously this step is modified by coupling to PEI at the N-terminus. In fact, coupling melittin via the C-terminus generates PEI conjugates which are toxic, destabilize the plasma membrane of cells, and do not mediate efficient transfection [181].

Another endosomal escape strategy uses listeriolysin O (LLO) which is a sulfhydryl-activated pore-forming protein from Listeria monocytogenes. To apply it as a membrane-disruptive agent with specificity for endosomes, LLO was conjugated through a reversible, endosome labile disulfide bond to the polycationic peptide protamine at a 1 : 1 molar ratio [182]. The LLO-S-S-protamine conjugate lacks pore-forming activity which, however, can be regained upon reduction. As evaluated in several cell culture systems, by incorporating increasing amounts of LLO-SS-protamine into DNA polyplexes, luciferase marker gene expression was enhanced. This endosomolytic system was superior to chloroquine-mediated transfection. No cytotoxicity was observed at the applied doses, in contrast to high cell lysis when free LLO was applied to cells together with protamine/DNA polyplexes.

Recently, an alternative novel technique was developed to improve endosomal release in a different way: the light-induced photochemical rupture of endocytic vesicles [41, 42, 183]. In this process cells are treated with amphiphilic photosensitizers followed by illumination. The photosensitizers localized in membranes of endocytic vesicles are activated by light, resulting in the destruction of endocytic membrane structures and releasing co-endocytosed polyplexes into the cell cytosol. This method termed "PCI, photochemical internalization" has already shown very encouraging results in improving polylysine or PEI polyplexes without or with targeting ligands [41, 43–45, 57].

Nuclear Import

Nuclear entry of non-viral vectors is another big hurdle which is currently only overcome in rapidly dividing cells; for example, transfection of non-dividing cells with lipoplexes or PEI polyplexes was several log units less effective compared to transfection of mitotic cells where the nuclear envelope has broken down [184, 185]. However, differences were observed with different polyplex formulations. A comparison of gene transfer properties of L-PEI and B-PEI revealed that within a few hours after in vitro transfection with L-PEI polyplexes, DNA particles are seen not only in the cytoplasm but even passing into the nucleus, whereas complexes with B-PEI were visible only in cytoplasmic structures with almost no DNA associated with the nucleus [55]. These data correlated with findings of lower cell-cycle dependence for L-PEI [22] and suggest nuclear uptake as the responsible event.

For polyplex formulations which exhibit strong cell cycle dependence of transfection, incorporation of biological nuclear localization signals [186] may be required when gene transfer into non-dividing cells is considered.

5
Bioresponsive Polymers—Towards Artificial Viruses

Current polyplexes are still very inefficient as compared to viral vectors. From this perspective, viruses might present ideal natural examples educating us how to further optimize polyplexes into "synthetic viruses" [10, 187]. One unique property of viruses is their dynamic manner in responding to the biological micro-environment. Similar to viruses, also polyplexes should alter their structure during the gene delivery process to make them most effective for the different subsequent gene delivery step. To obtain such a "chameleon"-like capability, bioresponsive polymers have to be incorporated that enable structural and functional changes triggered by the microenvironment, such as conformational changes or cleavage of chemical bonds. First examples of bioresponsive polyplex systems triggered by, for example, an acidic pH or a disulfide reducing environment are listed in Table 1.

Table 1 Examples of bioresponsive polyplex systems

	Action	Consequences	
Acidic endosomal pH as trigger			
Plank 1992, Mechtler 1997	pH-specific influenza peptide activated	endosomal escape	enhances gene transfer of pLL polyplexes
Rozema 2003	deprotection of succ-melittin	lytic activity recovered	endosomal escape
Murthy 2003	cleavage of acetal-linked PEG	endosomolytic polymer exposed	endosomal escape of antisense oligonucleotides
Walker 2004	cleavage of hydrazone-linked PEG	exposure of PEI polyplex positive surface charge	up to 100-fold enhanced gene transfer
Reducing environment as trigger			
Saito 2003	release of LLO from disulfide	lytic activity recovered	endosomal escape, increased gene expression
Trubetzkoy 1999	release of "cage" linkers	strong DNA binding is reduced within cell	intracellular accessibility of DNA for transcription
Read 2003, Carlisle 2004	reducible polycation (RPC) cleaved	DNA binding reduced within cell, hydrophilic shield removed	enhanced gene transfer of lipopolyplexes and shielded polyplexes
Miyata 2004	disulfide cross-linkers between PEG-pLL blocks cleaved	DNA binding reduced within cell, hydrophilic shield removed	increased gene expression

For instance, shielding and targeting molecules like PEG and receptor ligands are required only in the early extracellular steps of the delivery process. As indicated earlier (see Sect. 4.2), the introduction of shielding agents such as PEG is a double-edged sword: particles with a higher degree of PEG shield show longer circulation time and better accumulation at the target site, but lower gene expression activity in the target cells. Apparently, a stable, irreversible PEG shield hampers intracellular uptake processes. In an optimized virus-like polyplex, PEG shielding should be presented in a bio-responsive fashion. After entering the target cell and delivery into the endosomal vesicle, a polyplex should release the PEG shield, and the cationic surface of the polyplex should be re-exposed for efficient destabilization of the endosomal membrane. Such a dynamic change in character can be introduced into polyplexes by bioresponsive domains.

Walker et al. [133] made use of the acidic milieu of the endosomes and introduced bioresponsive PEG-polycation conjugates with pH-labile linkages (see Fig. 4). DNA particles shielded with these bio-reversible PEG conjugates

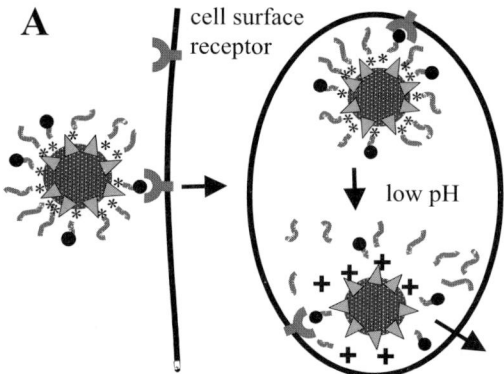

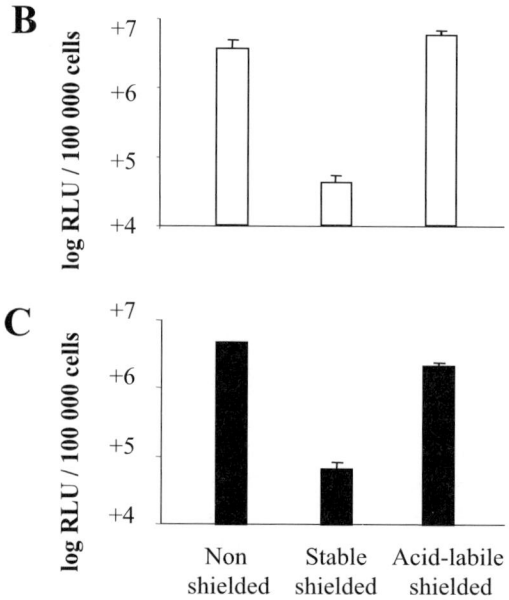

Fig. 4 Towards synthetic viruses. **A** Scheme illustrating receptor mediated uptake of transferrin-PEG-PEI polyplexes or EGF-PEG-PEI polyplexes into endosomal vesicles of K562 cells or HUH7 cells, respectively. The applied bioresponsive PEG-polycation conjugates are cleavable within the endosomal pH environment [133] facilitating subsequent escape of polyplexes from the endosome, resulting in strongly enhanced gene expression as compared with polyplexes containing the same amount of stable PEG-polycation conjugates. **B, C** Effect of stable or acid-labile PEG shield on polyplex transfection efficiency. **B** Transfection of K562 (*white bars*) cells using transferrin receptor-targeted polyplexes containing L-PEI 22 kDa for DNA condensation, plus Tf-PEG 2.4 kDa-PEI targeting conjugate, plus either unmodified PLL ("*non-shielded* polyplexes"), PLL-PEG 20 kDa ("*stable-shielded* polyplexes"), or PLL-HZN-PEG 20 kDa ("*acid-labile-shielded* polyplexes") conjugates. **C** Transfection of HUH7 cells (*black bars*) using EGF receptor-targeted polyplexes containing EGF-PEG 2.4 kDa-PEI plus other conjugates as indicated in **B**

(see Sect. 3.3.2) lose their PEG shield at endosomal pH and display up to 100-fold higher gene transfer activity as compared to polyplexes with the analogous stable PEG shield.

Apart from releasing a PEG shield, the acidic endosomal pH may also result in activation of membrane disrupting functions required within the endosome but not before. The acidification may activate acidic membrane-active peptides such as derived from influenza virus HA2 by pH-specific conformational changes [121, 151]. Alternatively, the low pH may trigger the removal of a masking group, such as described by Rozema et al. [188] where melittin is unmasked by cleavage of maleamate protective groups from melittin lysine residues, which recovers the lytic activity of melittin. Analogously, removal of an acetal-linked PEG molecule exposes a membrane-disruptive poly(propylacrylic acid) backbone [132].

The intracellular reducing environment may contribute to polyplex activation and disassembly by cleaving disulfide-bridged cationic carriers (compare Sect. 3.3.3). Triggered by the reducing environment, a lytic form of LLO was released from an inactive precursor, see Saito et al. [182]. In addition, several reports describe bioresponsive vectors that combine extracellular stability of DNA by polycationic disulfide-bond containing "cages" with rapid intracellular release of the DNA upon cleavage of these cages [144–146, 189]. For example, Carlisle et al. [189] coated DNA/PEI polyplexes containing 20% mercapto-modified PEI-SH with the hydrophilic shielding polymer pHPMA via either reducible disulfide or stable thioether bonds. Polyplexes with the disulfide linked coating show a higher activity than stable thioether coated complexes.

These examples demonstrate that incorporation of delivery functions that are presented in a bio-responsive fashion can strongly improve polyplex efficiency. It underlines the notion that developing polyplexes into virus-like supramolecular complexes, which alter their structure to cope best with the initial extracellular and subsequent intracellular gene delivery steps, is a promising direction to generate efficient synthetic gene therapy vectors.

6
Conclusions

Despite first results demonstrating that polyplexes can mediate gene therapeutic effects in animal models, these studies also demonstrated clear limitations. The systemic targeting efficiencies are by far not perfect; polyplex formulations often have significant toxic properties, and low in vivo gene transfer activity.

For clinical use, better defined, homogeneous and biocompatible systems will be necessary. Aspects for optimization will include a defined assembly into monodisperse particle populations preferably of small size [190]; methods for the purification of polyplexes [56] which remove potentially toxic free polymers and polyplex aggregates; and the generation of stabilized polyplexes with increased storage stability [92]. In addition, bio-degradable versions of polymers will have to be applied to reduce cellular toxicity, and to avoid long-term deposition of (non-degradable) polymers within the organism. They will also provide a bio-responsive, virus-like character required for improving extra- and intra-cellular delivery of the therapeutic nucleic acid.

Acknowledgements The contributions to our research by Carolin Fella, Greg Walker, Julia Fahrmeir, Katharina von Gersdorff, Manfred Ogris, Sabine Boeckle, Silke van der Piepen and Wolfgang Rödl are greatly appreciated. It has been an exceptional pleasure to work within such a fantastic team. We are very grateful to Dr. Greg Walker for the careful review of the manuscript. Many thanks to Olga Brück for skilful assistance in preparing the manuscript.

References

1. Duncan R (2003) The dawning era of polymer therapeutics. Nature Review 2:347–360
2. Berton M, Allemann E, Stein CA, Gurny R (1999) Eur J Pharm Sci 9:163
3. Truong L, Walsh SM, Schweibert E, Mao HQ, Guggino WB, August JT, Leong KW (1999) Arch Biochem Biophys 361:47
4. Mao HQ, Roy K, Troung L, Janes KA, Lin KY, Wang Y, August JT, Leong KW (2001) J Control Release 70:399
5. Cui Z, Mumper RJ (2002) Bioconjug Chem 13:1319
6. Felgner PL, Barenholz Y, Behr JP, Cheng SH, Cullis P, Huang L, Jessee JA, Seymour L, Szoka F, Thierry AR, Wagner E, Wu G (1997) Hum Gene Ther 8:511
7. De Smedt SC, Demeester J, Hennink WE (2000) Pharm Res 17:113
8. Brown MD, Schatzlein AG, Uchegbu IF (2001) Int J Pharm 229:1
9. Han S, Mahato RI, Sung YK, Kim SW (2000) Mol Ther 2:302
10. Wagner E (2004) Pharm Res 21:8
11. Vaheri A, Pagano JS (1965) Virology 27:434
12. Mislick KA, Baldeschwieler JD (1996) Proc Natl Acad Sci USA 93:12349
13. Goncalves C, Pichon C, Guerin B, Midoux P (2002) J Gene Med 4:271
14. Goncalves C, Mennesson E, Fuchs R, Gorvel JP, Midoux P, Pichon C (2004) Mol Ther 10:373
15. Kopatz I, Remy JS, Behr JP (2004) J Gene Med 6:769
16. Ruponen M, Honkakoski P, Tammi M, Urtti A (2004) J Gene Med 6:405
17. Wu GY, Wu CH (1987) J Biol Chem 262:4429
18. Wagner E, Zenke M, Cotten M, Beug H, Birnstiel ML (1990) Proc Natl Acad Sci USA 87:3410
19. Curiel DT, Agarwal S, Wagner E, Cotten M (1991) Proc Natl Acad Sci USA 88:8850
20. Lukacs GL, Haggie P, Seksek O, Lechardeur D, Freedman N, Verkman AS (2000) J iol Chem 275:1625
21. Ludtke JJ, Zhang G, Sebestyén MG, Wolff JA (1999) J Cell Sci 112:2033
22. Brunner S, Furtbauer E, Sauer T, Kursa M, Wagner E (2002) Mol Ther 5:80
23. Schaffer DV, Fidelman NA, Dan N, Lauffenburger DA (2000) Biotechnol Bioeng 67:598
24. Plank C, Tang MX, Wolfe AR, Szoka FC Jr (1999) Hum Gene Ther 10:319

25. Parker AL, Oupicky D, Dash PR, Seymour LW (2002) Anal Biochem 302:75
26. Reineke TM, Davis ME (2003) Bioconjug Chem 14:255
27. Reineke TM, Davis ME (2003) Bioconjug Chem 14:247
28. Chen QR, Zhang L, Luther PW, Mixson AJ (2002) Nucleic Acids Res 30:1338
29. Zauner W, Ogris M, Wagner E (1998) Advanced Drug Delivery Reviews 30(1-3):97
30. Wang J, Gao SJ, Zhang PC, Wang S, Mao HQ, Leong KW (2004) Gene Ther 11:1001
31. Wagner E, Cotten M, Foisner R, Birnstiel ML (1991) Proc Natl Acad Sci USA 88:4255
32. Wu GY, Wu CH (1988) J Biol Chem 262:14621
33. Chowdhury NR, Wu CH, Wu GY, Yerneni PC, Bommineni VR, Chowdhury JR (1993) J Biol Chem 268:11265
34. Wu GY, Wilson JM, Shalaby F, Grossman M, Shafritz DA, Wu CH (1991) J Biol Chem 266:14338
35. Wilson JM, Grossman M, Wu CH, Chowdhury NR, Wu GY, Chowdhury JR (1992) J Biol Chem 267:963
36. Perales JC, Ferkol T, Beegen H, Ratnoff OD, Hanson RW (1994) Proc Natl Acad Sci USA 91:4086
37. Ferkol T, Perales JC, Eckman E, Kaetzel CS, Hanson RW, Davis PB (1995) J Clin Invest 95:493
38. Ziady AG, Ferkol T, Dawson DV, Perlmutter DH, Davis PB (1999) J Biol Chem 274:4908
39. Cotten M, Langle R, Kirlappos H, Wagner E, Mechtler K, Zenke M, Beug H, Birnstiel ML (1990) Proc Natl Acad Sci USA 87:4033
40. Zauner W, Blaas D, Kuechler E, Wagner E (1995) J Virol 69:1085
41. Hogset A, Prasmickaite L, Tjelle TE, Berg K (2000) Hum Gene Ther 11:869
42. Prasmickaite L, Hogset A, Selbo PK, Engesaeter BO, Hellum M, Berg K (2002) Br J Cancer 86:652
43. Prasmickaite L, Hogset A, Tjelle TE, Olsen VM, Berg K (2000) J Gene Med 2:477
44. Prasmickaite L, Hogset A, Berg K (2001) Photochem Photobiol 73:388
45. Prasmickaite L, Hogset A, Berg K (2002) Biochim Biophys Acta 1570:210
46. Boussif O, Lezoualc'h F, Zanta MA, Mergny MD, Scherman D, Demeneix B, Behr JP (1995) Proc Natl Acad Sci USA 92:7297
47. Zou SM, Erbacher P, Remy JS, Behr JP (2000) J Gene Med 2:128
48. Sonawane ND, Szoka FC Jr, Verkman AS (2003) J Biol Chem 278:44826
49. Goula D, Becker N, Lemkine GF, Normandie P, Rodrigues J, Mantero S, Levi G, Demeneix BA (2000) Gene Ther 7:499
50. Chollet P, Favrot MC, Hurbin A, Coll JL (2002) J Gene Med 4:84
51. Kircheis R, Schuller S, Brunner S, Heider K, Zauner W, Wagner E (1999) J Gene Med 1:111
52. Rudolph C, Lausier J, Naundorf S, Muller RH, Rosenecker J (2000) J Gene Med 2:269
53. Gautam A, Waldrep JC, Densmore CL, Koshkina N, Melton S, Roberts L, Gilbert B, Knight V (2002) Gene Ther 9:353
54. Ogris M, Steinlein P, Kursa M, Mechtler K, Kircheis R, Wagner E (1998) Gene Ther 5:1425
55. Wightman L, Kircheis R, Rossler V, Carotta S, Ruzicka R, Kursa M, Wagner E (2001) J Gene Med 3:362
56. Boeckle S, von Gersdorff K, van der Piepen S, Culmsee C, Wagner E, Ogris M (2004) J Gene Med 6:1102
57. Kloeckner J, Prasmickaite L, Hogset A, Berg K, Wagner E (2004) J Drug Targeting 12:205
58. Haensler J, Szoka FC Jr (1993) Bioconjug Chem 4:372

59. Tang MX, Szoka FC (1997) Gene Ther 4:823
60. Kukowska-Latallo JF, Bielinska AU, Johnson J, Spindler R, Tomalia DA, Baker JR Jr (1996) Proc Natl Acad Sci USA 93:4897
61. Kukowska-Latallo JF, Raczka E, Quintana A, Chen C, Rymaszewski M, Baker JR Jr (2000) Hum Gene Ther 11:1385
62. Manunta M, Tan PH, Sagoo P, Kashefi K, George AJ (2004) Nucleic Acids Res 32:2730
63. Mansouri S, Lavigne P, Corsi K, Benderdour M, Beaumont E, Fernandes JC (2004) Eur J Pharm Biopharm 57:1
64. Erbacher P, Zou S, Bettinger T, Steffan AM, Remy JS (1998) Pharm Res 15:1332
65. Park IK, Kim TH, Park YH, Shin BA, Choi ES, Chowdhury EH, Akaike T, Cho CS (2001) J Control Release 76:349
66. Thanou M, Florea BI, Geldof M, Junginger HE, Borchard G (2002) Biomaterials 23:153
67. Sato T, Ishii T, Okahata Y (2001) Biomaterials 22:2075
68. Kim TH, Ihm JE, Choi YJ, Nah JW, Cho CS (2003) J Control Release 93:389
69. Cherng JY, van de WP, Talsma H, Crommelin DJ, Hennink WE (1996) Pharm Res 13:1038
70. van de WP, Cherng JY, Talsma H, Crommelin DJ, Hennink WE (1998) J Control Release 53:145
71. van de WP, Schuurmans-Nieuwenbroek NM, Hennink WE, Storm G (1999) J Gene Med 1:156
72. Verbaan FJ, Oussoren C, van Dam IM, Takakura Y, Hashida M, Crommelin DJ, Hennink WE, Storm G (2001) Int J Pharm 214:99
73. Funhoff AM, van Nostrum CF, Koning GA, Schuurmans-Nieuwenbroek NM, Crommelin DJ, Hennink WE (2004) Biomacromolecules 5:32
74. Hwang SJ, Bellocq NC, Davis ME (2001) Bioconjug Chem 12:280
75. Bellocq NC, Pun SH, Jensen GS, Davis ME (2003) Bioconjug Chem 14:1122
76. Cryan SA, Holohan A, Donohue R, Darcy R, O'Driscoll CM (2004) Eur J Pharm Sci 21:625
77. Arima H, Kihara F, Hirayama F, Uekama K (2001) Bioconjug Chem 12:476
78. Kihara F, Arima H, Tsutsumi T, Hirayama F, Uekama K (2003) Bioconjug Chem 14:342
79. Pun SH, Bellocq NC, Liu A, Jensen G, Machemer T, Quijano E, Schluep T, Wen S, Engler H, Heidel J, Davis ME (2004) Bioconjug Chem 15:831
80. Kabanov AV, Batrakova EV, Alakhov VY (2002) J Control Release 82:189
81. Lemieux P, Guerin N, Paradis G, Proulx R, Chistyakova L, Kabanov A, Alakhov V (2000) Gene Ther 7:986
82. Nguyen HK, Lemieux P, Vinogradov SV, Gebhart CL, Guérin N, Paradis G, Bronich TK, Alakhov VY, Kabanov AV (2000) Gene Ther 7:126
83. Belenkov AI, Alakhov VY, Kabanov AV, Vinogradov SV, Panasci LC, Monia BP, Chow TY (2004) Gene Ther 22:1665
84. Li Z, Huang L (2004) J Control Release 98:437
85. Bello Roufai M, Midoux P (2001) Bioconjug Chem 12:92
86. Putnam D, Gentry CA, Pack DW, Langer R (2001) Proc Natl Acad Sci USA 98:1200
87. Kichler A (2004) J Gene Med 6:S3
88. Ogris M, Brunner S, Schueller S, Kircheis R, Wagner E (1999) Gene Ther 6:595
89. Petersen H, Fechner PM, Martin AL, Kunath K, Stolnik S, Roberts CJ, Fischer D, Davies MC, Kissel T (2002) Bioconjug Chem 13:845
90. Kichler A, Chillon M, Leborgne C, Danos O, Frisch B (2002) J Control Release 81:379
91. Kursa M, Walker GF, Roessler V, Ogris M, Roedl W, Kircheis R, Wagner E (2003) Bioconjug Chem 14:222

92. Ogris M, Walker G, Blessing T, Kircheis R, Wolschek M, Wagner E (2003) J Control Release 91:173
93. Hong JW, Park JH, Huh KM, Chung H, Kwon IC, Jeong SY (2004) J Control Release 99:167
94. Tseng WC, Jong CM (2003) Biomacromolecules 4:1277
95. Brissault B, Kichler A, Guis C, Leborgne C, Danos O, Cheradame H (2003) Bioconjug Chem 14:581
96. Jeong JH, Song SH, Lim DW, Lee H, Park TG (2001) J Control Release 73:391
97. Thomas M, Klibanov AM (2002) Proc Natl Acad Sci USA 99:14640
98. Fischer D, von Harpe A, Kunath K, Petersen H, Li Y, Kissel T (2002) Bioconjug Chem 13:1124
99. Brownlie A, Uchegbu IF, Schatzlein AG (2004) Int J Pharm 274:41
100. Forrest ML, Meister GE, Koerber JT, Pack DW (2004) Pharm Res 21:365
101. Wang DA, Narang AS, Kotb M, Gaber AO, Miller DD, Kim SW, Mahato RI (2002) Biomacromolecules 3:1197
102. Furgeson DY, Chan WS, Yockman JW, Kim SW (2003) Bioconjug Chem 14:840
103. Sochanik A, Cichon T, Makselon M, Strozyk M, Smolarczyk R, Jazowiecka-Rakus J, Szala S (2004) Acta Biochim Pol 51:693
104. Choi JS, Nam K, Park JY, Kim JB, Lee JK, Park JS (2004) J Control Release 99:445
105. Ohsaki M, Okuda T, Wada A, Hirayama T, Niidome T, Aoyagi H (2002) Bioconjug Chem 13:510
106. Okuda T, Sugiyama A, Niidome T, Aoyagi H (2004) Biomaterials 25:537
107. Kawano T, Okuda T, Aoyagi H, Niidome T (2004) J Control Release 99:329
108. Zinselmeyer BH, Mackay SP, Schatzlein AG, Uchegbu IF (2002) Pharm Res 19:960
109. Uchegbu IF, Dufes C, Elouzi A, Schatzlein AG (2004) Proceed 12th Int Pharm Techn Symp 2004:11
110. Banerjee P, Reichardt W, Weissleder R, Bogdanov A Jr (2004) Bioconjug Chem 15:960
111. Kramer M, Stumbe JF, Grimm G, Kaufmann B, Kruger U, Weber M, Haag R (2004) Chembiochem 5:1081
112. Fischer D, Bieber T, Li Y, Elsasser HP, Kissel T (1999) Pharm Res 16:1273
113. Ahn CH, Chae SY, Bae YH, Kim SW (2004) J Control Release 97:567
114. Lim YB, Han SO, Kong HU, Lee Y, Park JS, Jeong B, Kim SW (2000) Pharm Res 17:811
115. Lim Y, Choi YH, Park J (1999) J Am Chem Soc 121:5633
116. Anderson DG, Lynn DM, Langer R (2003) Angew Chem Int Ed Engl 42:3153
117. Akinc A, Anderson DG, Lynn DM, Langer R (2003) Bioconjug Chem 14:979
118. Lim YB, Kim SM, Suh H, Park JS (2002) Bioconjug Chem 13:952
119. Oster CG, Wittmar M, Unger F, Barbu-Tudoran L, Schaper AK, Kissel T (2004) Pharm Res 21:927
120. Funhoff AM, van Nostrum CF, Janssen AP, Fens MH, Crommelin DJ, Hennink WE (2004) Pharm Res 21:170
121. Mechtler K, Wagner E (1997) New J Chem 21:105
122. Ahn CH, Chae SY, Bae YH, Kim SW (2002) J Control Release 80:273
123. Petersen H, Merdan T, Kunath K, Fischer D, Kissel T (2002) Bioconjug Chem 13:812
124. Forrest ML, Koerber JT, Pack DW (2003) Bioconjug Chem 14:934
125. Zhao Z, Wang J, Mao HQ, Leong KW (2003) Adv Drug Deliv Rev 55:483
126. Wang J, Mao HQ, Leong KW (2001) J Am Chem Soc 123:9480
127. Wang J, Zhang PC, Mao HQ, Leong KW (2002) Gene Ther 9:1254
128. Wang J, Zhang PC, Lu HF, Ma N, Wang S, Mao HQ, Leong KW (2002) J Control Release 83:157

129. Asokan A, Cho M (2002) J Pharm Sci 91:903
130. Wang C, Ge Q, Ting D, Nguyen D, Shen HR, Chen J, Eisen HN, Heller J, Langer R, Putnam D (2004) Nat Mater 3:190
131. Murthy N, Campbell J, Fausto N, Hoffman AS, Stayton PS (2003) J Control Release 89:365
132. Lackey CA, Press OW, Hoffman AS, Stayton PS (2002) Bioconjug Chem 13:996
133. Walker GF, Fella C, Pelisek J, Fahrmeir J, Boeckle S, Ogris M, Wagner E (2005) Mol Ther 11:418
134. Saito G, Swanson JA, Lee KD (2003) Adv Drug Deliv Rev 55:199
135. Kakizawa Y, Harada A, Kataoka K (1999) J Am Chem Soc 121:11247
136. Kakizawa Y, Harada A, Kataoka K (2001) Biomacromolecules 2:491
137. Collins DS, Unanue ER, Harding CV (1991) J Immunol 147:4054
138. McKenzie DL, Smiley E, Kwok KY, Rice KG (2000) Bioconjug Chem 11:901
139. McKenzie DL, Kwok KY, Rice KG (2000) J Biol Chem 275:9970
140. Park Y, Kwok KY, Boukarim C, Rice KG (2002) Bioconjug Chem 13:232
141. Kwok KY, Park Y, Yang Y, McKenzie DL, Liu Y, Rice KG (2003) J Pharm Sci 92:1174
142. Oupicky D, Carlisle RC, Seymour LW (2001) Gene Ther 8:713
143. Oupicky D, Parker AL, Seymour LW (2002) J Am Chem Soc 124:8
144. Read ML, Bremner KH, Oupicky D, Green NK, Searle PF, Seymour LW (2003) J Gene Med 5:232
145. Trubetskoy VS, Loomis A, Slattum PM, Hagstrom JE, Budker VG, Wolff JA (1999) Bioconjug Chem 10:624
146. Miyata K, Kakizawa Y, Nishiyama N, Harada A, Yamasaki Y, Koyama H, Kataoka K (2004) J Am Chem Soc 126:2355
147. Pichon C, LeCam E, Guerin B, Coulaud D, Delain E, Midoux P (2002) Bioconjug Chem 13:76
148. Gosselin MA, Guo W, Lee RJ (2001) Bioconjug Chem 12:989
149. Guo W, Lee RJ (1999) AAPS Pharmsci 1:Article 19
150. Mislick KA, Baldeschwieler JD, Kayyem JF, Meade TJ (1995) Bioconjug Chem 6:512
151. Plank C, Zatloukal K, Cotten M, Mechtler K, Wagner E (1992) Bioconjug Chem 3:533
152. Nishikawa M, Yamauchi M, Morimoto K, Ishida E, Takakura Y, Hashida M (2000) Gene Ther 7:548
153. Kim TG, Kang SY, Kang JH, Cho MY, Kim JI, Kim SH, Kim JS (2004) Bioconjug Chem 15:326
154. Wolschek MF, Thallinger C, Kursa M, Rossler V, Allen M, Lichtenberger C, Kircheis R, Lucas T, Willheim M, Reinisch W, Gangl A, Wagner E, Jansen B (2002) Hepatology 36:1106
155. Sosnowski BA, Gonzalez AM, Chandler LA, Buechler YJ, Pierce GF, Baird A (1996) J Biol Chem 271:33647
156. Chiu SJ, Ueno NT, Lee RJ (2004) J Control Release 97:357
157. Buschle M, Cotten M, Kirlappos H, Mechtler K, Schaffner G, Zauner W, Birnstiel ML, Wagner E (1995) Hum Gene Ther 6:753
158. Merdan T, Callahan J, Petersen H, Kunath K, Bakowsky U, Kopeckova P, Kissel T, Kopecek J (2003) Bioconjug Chem 14:989
159. Li S, Tan Y, Viroonchatapan E, Pitt BR, Huang L (2000) Am J Physiol Lung Cell Mol Physiol 278:504
160. Schatzlein AG (2003) J Biomed Biotechnol 2003:149
161. Wagner E, Culmsee C, Boeckle S (2005) Targeting of Polyplexes: Towards Synthetic Virus Vector Systems. In: Huang L, Hung MC, Wagner E (eds) Nonviral Vectors for Gene Therapy, Second Edition (Advances in Genetics, vol 53)

162. Kircheis R, Wightman L, Schreiber A, Robitza B, Rossler V, Kursa M, Wagner E (2001) Gene Ther 8:28
163. Plank C, Mechtler K, Szoka FJ, Wagner E (1996) Hum Gene Ther 7:1437
164. Verbaan FJ, Oussoren C, Snel CJ, Crommelin DJ, Hennink WE, Storm G (2004) J Gene Med 6:64
165. Fisher KD, Ulbrich K, Subr V, Ward CM, Mautner V, Blakey D, Seymour LW (2000) Gene Ther 7:1337
166. Kircheis R, Ostermann E, Wolschek MF, Lichtenberger C, Magin-Lachmann C, Wightman L, Kursa M, Wagner E (2002) Cancer Gene Ther 9:673
167. Wagner E, Zatloukal K, Cotten M, Kirlappos H, Mechtler K, Curiel DT, Birnstiel ML (1992) Proc Natl Acad Sci USA 89:6099
168. Rejman J, Oberle V, Zuhorn IS, Hoekstra D (2004) Biochem J 377:159
169. Plank C, Zauner W, Wagner E (1998) Adv Drug Deliv Rev 34:21
170. Boeckle S, Wagner E, Ogris M (2002) Transmembrane Targeting of DNA with Membrane Active Peptides. In: Muzykantov VR, Torchilin VP (eds) Biomedical Aspects of Drug Targeting. Kluwer Academic Publishers, Boston/Dordrecht/London Chap 23, p 441
171. Cristiano RJ, Smith LC, Kay MA, Brinkley BR, Woo SL (1993) Proc Natl Acad Sci USA 90:11548
172. Wu GY, Zhan P, Sze LL, Rosenberg AR, Wu CH (1994) J Biol Chem 269:11542
173. Fisher KJ, Wilson JM (1994) Biochem J 299:49
174. Fender P, Ruigrok RW, Gout E, Buffet S, Chroboczek J (1997) Nat Biotechnol 15:52
175. Gottschalk S, Tweten RK, Smith LC, Woo SL (1995) Gene Ther 2:498
176. Fisher KJ, Wilson JM (1997) Biochem J 321:49
177. Gottschalk S, Sparrow JT, Hauer J, Mims MP, Leland FE, Woo SL, Smith LC (1996) Gene Ther 3:48
178. Ogris M, Carlisle RC, Bettinger T, Seymour LW (2001) J Biol Chem 276:47550
179. Kichler A, Mechtler K, Behr JP, Wagner E (1997) Bioconjug Chem 8:213
180. Bettinger T, Carlisle RC, Read ML, Ogris M, Seymour LW (2001) Nucleic Acids Res 29:3882
181. Boeckle S, Ogris M, Wagner E (2005) J Gene Med 7:1335
182. Saito G, Amidon GL, Lee KD (2003) Gene Ther 10:72
183. Berg K, Selbo PK, Prasmickaite L, Tjelle TE, Sandvig K, Moan J, Gaudernack G, Fodstad O, Kjolsrud S, Anholt H, Rodal GH, Rodal SK, Hogset A (1999) Cancer Res 59:1180
184. Zabner J, Fasbender AJ, Moninger T, Poellinger KA, Welsh MJ (1995) J Biol Chem 270:18997
185. Brunner S, Sauer T, Carotta S, Cotten M, Saltik M, Wagner E (2000) Gene Ther 7:401
186. Sebestyen MG, Ludtke JJ, Bassik MC, Zhang G, Budker V, Lukhtanov EA, Hagstrom JE, Wolff JA (1998) Nat Biotechnol 16:80
187. Zuber G, Dauty E, Nothisen M, Belguise P, Behr JP (2001) Adv Drug Deliv Rev 52:245
188. Rozema DB, Ekena K, Lewis DL, Loomis AG, Wolff JA (2003) Bioconjug Chem 14:51
189. Carlisle RC, Etrych T, Briggs SS, Preece JA, Ulbrich K, Seymour LW (2004) J Gene Med 6:337
190. Blessing QT, Remy JS, Behr JP (1998) Proc Natl Acad Sci USA 95:1427

Author Index Volumes 101–192

Author Index Volumes 1–100 see Volume 100

de Abajo, J. and *de la Campa, J. G.*: Processable Aromatic Polyimides. Vol. 140, pp. 23–60.
Abe, A., Furuya, H., Zhou, Z., Hiejima, T. and *Kobayashi, Y.*: Stepwise Phase Transitions of Chain Molecules: Crystallization/Melting via a Nematic Liquid-Crystalline Phase. Vol. 181, pp. 121–152.
Abetz, V. and *Simon, P. F. W.*: Phase Behaviour and Morphologies of Block Copolymers. Vol. 189, pp. 125–212.
Abetz, V. see Förster, S.: Vol. 166, pp. 173–210.
Adolf, D. B. see Ediger, M. D.: Vol. 116, pp. 73–110.
Aharoni, S. M. and *Edwards, S. F.*: Rigid Polymer Networks. Vol. 118, pp. 1–231.
Albertsson, A.-C. and *Varma, I. K.*: Aliphatic Polyesters: Synthesis, Properties and Applications. Vol. 157, pp. 99–138.
Albertsson, A.-C. see Edlund, U.: Vol. 157, pp. 53–98.
Albertsson, A.-C. see Söderqvist Lindblad, M.: Vol. 157, pp. 139–161.
Albertsson, A.-C. see Stridsberg, K. M.: Vol. 157, pp. 27–51.
Albertsson, A.-C. see Al-Malaika, S.: Vol. 169, pp. 177–199.
Allegra, G. and *Meille, S. V.*: Pre-Crystalline, High-Entropy Aggregates: A Role in Polymer Crystallization? Vol. 191, pp. 87–135.
Al-Malaika, S.: Perspectives in Stabilisation of Polyolefins. Vol. 169, pp. 121–150.
Altstädt, V.: The Influence of Molecular Variables on Fatigue Resistance in Stress Cracking Environments. Vol. 188, pp. 105–152.
Améduri, B., Boutevin, B. and *Gramain, P.*: Synthesis of Block Copolymers by Radical Polymerization and Telomerization. Vol. 127, pp. 87–142.
Améduri, B. and *Boutevin, B.*: Synthesis and Properties of Fluorinated Telechelic Monodispersed Compounds. Vol. 102, pp. 133–170.
Ameduri, B. see Taguet, A.: Vol. 184, pp. 127–211.
Amir, R. J. and *Shabat, D.*: Domino Dendrimers. Vol. 192, pp. 59–94.
Amselem, S. see Domb, A. J.: Vol. 107, pp. 93–142.
Anantawaraskul, S., Soares, J. B. P. and *Wood-Adams, P. M.*: Fractionation of Semicrystalline Polymers by Crystallization Analysis Fractionation and Temperature Rising Elution Fractionation. Vol. 182, pp. 1–54.
Andrady, A. L.: Wavelenght Sensitivity in Polymer Photodegradation. Vol. 128, pp. 47–94.
Andreis, M. and *Koenig, J. L.*: Application of Nitrogen-15 NMR to Polymers. Vol. 124, pp. 191–238.
Angiolini, L. see Carlini, C.: Vol. 123, pp. 127–214.
Anjum, N. see Gupta, B.: Vol. 162, pp. 37–63.
Anseth, K. S., Newman, S. M. and *Bowman, C. N.*: Polymeric Dental Composites: Properties and Reaction Behavior of Multimethacrylate Dental Restorations. Vol. 122, pp. 177–218.
Antonietti, M. see Cölfen, H.: Vol. 150, pp. 67–187.
Aoki, H. see Ito, S.: Vol. 182, pp. 131–170.

Armitage, B. A. see O'Brien, D. F.: Vol. 126, pp. 53–58.
Arnal, M. L. see Müller, A. J.: Vol. 190, pp. 1–63.
Arndt, M. see Kaminski, W.: Vol. 127, pp. 143–187.
Arnold, A. and *Holm, C.*: Efficient Methods to Compute Long-Range Interactions for Soft Matter Systems. Vol. 185, pp. 59–109.
Arnold Jr., F. E. and *Arnold, F. E.*: Rigid-Rod Polymers and Molecular Composites. Vol. 117, pp. 257–296.
Arora, M. see Kumar, M. N. V. R.: Vol. 160, pp. 45–118.
Arshady, R.: Polymer Synthesis via Activated Esters: A New Dimension of Creativity in Macromolecular Chemistry. Vol. 111, pp. 1–42.
Auer, S. and *Frenkel, D.*: Numerical Simulation of Crystal Nucleation in Colloids. Vol. 173, pp. 149–208.
Auriemma, F., de Rosa, C. and *Corradini, P.*: Solid Mesophases in Semicrystalline Polymers: Structural Analysis by Diffraction Techniques. Vol. 181, pp. 1–74.

Bahar, I., Erman, B. and *Monnerie, L.*: Effect of Molecular Structure on Local Chain Dynamics: Analytical Approaches and Computational Methods. Vol. 116, pp. 145–206.
Baietto-Dubourg, M. C. see Chateauminois, A.: Vol. 188, pp. 153–193.
Ballauff, M. see Dingenouts, N.: Vol. 144, pp. 1–48.
Ballauff, M. see Holm, C.: Vol. 166, pp. 1–27.
Ballauff, M. see Rühe, J.: Vol. 165, pp. 79–150.
Balsamo, V. see Müller, A. J.: Vol. 190, pp. 1–63.
Baltá-Calleja, F. J., González Arche, A., Ezquerra, T. A., Santa Cruz, C., Batallón, F., Frick, B. and *López Cabarcos, E.*: Structure and Properties of Ferroelectric Copolymers of Poly(vinylidene) Fluoride. Vol. 108, pp. 1–48.
Baltussen, J. J. M. see Northolt, M. G.: Vol. 178, pp. 1–108.
Barnes, M. D. see Otaigbe, J. U.: Vol. 154, pp. 1–86.
Barsett, H. see Paulsen, S. B.: Vol. 186, pp. 69–101.
Barshtein, G. R. and *Sabsai, O. Y.*: Compositions with Mineralorganic Fillers. Vol. 101, pp. 1–28.
Barton, J. see Hunkeler, D.: Vol. 112, pp. 115–134.
Baschnagel, J., Binder, K., Doruker, P., Gusev, A. A., Hahn, O., Kremer, K., Mattice, W. L., Müller-Plathe, F., Murat, M., Paul, W., Santos, S., Sutter, U. W. and *Tries, V.*: Bridging the Gap Between Atomistic and Coarse-Grained Models of Polymers: Status and Perspectives. Vol. 152, pp. 41–156.
Bassett, D. C.: On the Role of the Hexagonal Phase in the Crystallization of Polyethylene. Vol. 180, pp. 1–16.
Batallán, F. see Baltá-Calleja, F. J.: Vol. 108, pp. 1–48.
Batog, A. E., Pet'ko, I. P. and *Penczek, P.*: Aliphatic-Cycloaliphatic Epoxy Compounds and Polymers. Vol. 144, pp. 49–114.
Baughman, T. W. and *Wagener, K. B.*: Recent Advances in ADMET Polymerization. Vol. 176, pp. 1–42.
Becker, O. and *Simon, G. P.*: Epoxy Layered Silicate Nanocomposites. Vol. 179, pp. 29–82.
Bell, C. L. and *Peppas, N. A.*: Biomedical Membranes from Hydrogels and Interpolymer Complexes. Vol. 122, pp. 125–176.
Bellon-Maurel, A. see Calmon-Decriaud, A.: Vol. 135, pp. 207–226.
Bennett, D. E. see O'Brien, D. F.: Vol. 126, pp. 53–84.
Berry, G. C.: Static and Dynamic Light Scattering on Moderately Concentraded Solutions: Isotropic Solutions of Flexible and Rodlike Chains and Nematic Solutions of Rodlike Chains. Vol. 114, pp. 233–290.

Bershtein, V. A. and *Ryzhov, V. A.*: Far Infrared Spectroscopy of Polymers. Vol. 114, pp. 43–122.
Bhargava, R., Wang, S.-Q. and *Koenig, J. L*: FTIR Microspectroscopy of Polymeric Systems. Vol. 163, pp. 137–191.
Biesalski, M. see Rühe, J.: Vol. 165, pp. 79–150.
Bigg, D. M.: Thermal Conductivity of Heterophase Polymer Compositions. Vol. 119, pp. 1–30.
Binder, K.: Phase Transitions in Polymer Blends and Block Copolymer Melts: Some Recent Developments. Vol. 112, pp. 115–134.
Binder, K.: Phase Transitions of Polymer Blends and Block Copolymer Melts in Thin Films. Vol. 138, pp. 1–90.
Binder, K. see Baschnagel, J.: Vol. 152, pp. 41–156.
Binder, K., Müller, M., Virnau, P. and *González MacDowell, L.*: Polymer+Solvent Systems: Phase Diagrams, Interface Free Energies, and Nucleation. Vol. 173, pp. 1–104.
Bird, R. B. see Curtiss, C. F.: Vol. 125, pp. 1–102.
Biswas, M. and *Mukherjee, A.*: Synthesis and Evaluation of Metal-Containing Polymers. Vol. 115, pp. 89–124.
Biswas, M. and *Sinha Ray, S.*: Recent Progress in Synthesis and Evaluation of Polymer-Montmorillonite Nanocomposites. Vol. 155, pp. 167–221.
Blankenburg, L. see Klemm, E.: Vol. 177, pp. 53–90.
Blumen, A. see Gurtovenko, A. A.: Vol. 182, pp. 171–282.
Bogdal, D., Penczek, P., Pielichowski, J. and *Prociak, A.*: Microwave Assisted Synthesis, Crosslinking, and Processing of Polymeric Materials. Vol. 163, pp. 193–263.
Bohrisch, J., Eisenbach, C. D., Jaeger, W., Mori, H., Müller, A. H. E., Rehahn, M., Schaller, C., Traser, S. and *Wittmeyer, P.*: New Polyelectrolyte Architectures. Vol. 165, pp. 1–41.
Bolze, J. see Dingenouts, N.: Vol. 144, pp. 1–48.
Bosshard, C.: see Gubler, U.: Vol. 158, pp. 123–190.
Boutevin, B. and *Robin, J. J.*: Synthesis and Properties of Fluorinated Diols. Vol. 102, pp. 105–132.
Boutevin, B. see Améduri, B.: Vol. 102, pp. 133–170.
Boutevin, B. see Améduri, B.: Vol. 127, pp. 87–142.
Boutevin, B. see Guida-Pietrasanta, F.: Vol. 179, pp. 1–27.
Boutevin, B. see Taguet, A.: Vol. 184, pp. 127–211.
Bowman, C. N. see Anseth, K. S.: Vol. 122, pp. 177–218.
Boyd, R. H.: Prediction of Polymer Crystal Structures and Properties. Vol. 116, pp. 1–26.
Bracco, S. see Sozzani, P.: Vol. 181, pp. 153–177.
Briber, R. M. see Hedrick, J. L.: Vol. 141, pp. 1–44.
Bronnikov, S. V., Vettegren, V. I. and *Frenkel, S. Y.*: Kinetics of Deformation and Relaxation in Highly Oriented Polymers. Vol. 125, pp. 103–146.
Brown, H. R. see Creton, C.: Vol. 156, pp. 53–135.
Bruza, K. J. see Kirchhoff, R. A.: Vol. 117, pp. 1–66.
Buchmeiser, M. R.: Regioselective Polymerization of 1-Alkynes and Stereoselective Cyclopolymerization of a, w-Heptadiynes. Vol. 176, pp. 89–119.
Budkowski, A.: Interfacial Phenomena in Thin Polymer Films: Phase Coexistence and Segregation. Vol. 148, pp. 1–112.
Bunz, U. H. F.: Synthesis and Structure of PAEs. Vol. 177, pp. 1–52.
Burban, J. H. see Cussler, E. L.: Vol. 110, pp. 67–80.
Burchard, W.: Solution Properties of Branched Macromolecules. Vol. 143, pp. 113–194.
Butté, A. see Schork, F. J.: Vol. 175, pp. 129–255.

Calmon-Decriaud, A., Bellon-Maurel, V., Silvestre, F.: Standard Methods for Testing the Aerobic Biodegradation of Polymeric Materials. Vol. 135, pp. 207–226.
Cameron, N. R. and *Sherrington, D. C.*: High Internal Phase Emulsions (HIPEs)-Structure, Properties and Use in Polymer Preparation. Vol. 126, pp. 163–214.
de la Campa, J. G. see de Abajo, J.: Vol. 140, pp. 23–60.
Candau, F. see Hunkeler, D.: Vol. 112, pp. 115–134.
Canelas, D. A. and *DeSimone, J. M.*: Polymerizations in Liquid and Supercritical Carbon Dioxide. Vol. 133, pp. 103–140.
Canva, M. and *Stegeman, G. I.*: Quadratic Parametric Interactions in Organic Waveguides. Vol. 158, pp. 87–121.
Capek, I.: Kinetics of the Free-Radical Emulsion Polymerization of Vinyl Chloride. Vol. 120, pp. 135–206.
Capek, I.: Radical Polymerization of Polyoxyethylene Macromonomers in Disperse Systems. Vol. 145, pp. 1–56.
Capek, I. and *Chern, C.-S.*: Radical Polymerization in Direct Mini-Emulsion Systems. Vol. 155, pp. 101–166.
Cappella, B. see Munz, M.: Vol. 164, pp. 87–210.
Carlesso, G. see Prokop, A.: Vol. 160, pp. 119–174.
Carlini, C. and *Angiolini, L.*: Polymers as Free Radical Photoinitiators. Vol. 123, pp. 127–214.
Carter, K. R. see Hedrick, J. L.: Vol. 141, pp. 1–44.
Casas-Vazquez, J. see Jou, D.: Vol. 120, pp. 207–266.
Chan, C.-M. and *Li, L.*: Direct Observation of the Growth of Lamellae and Spherulites by AFM. Vol. 188, pp. 1–41.
Chandrasekhar, V.: Polymer Solid Electrolytes: Synthesis and Structure. Vol. 135, pp. 139–206.
Chang, J. Y. see Han, M. J.: Vol. 153, pp. 1–36.
Chang, T.: Recent Advances in Liquid Chromatography Analysis of Synthetic Polymers. Vol. 163, pp. 1–60.
Charleux, B. and *Faust, R.*: Synthesis of Branched Polymers by Cationic Polymerization. Vol. 142, pp. 1–70.
Chateauminois, A. and *Baietto-Dubourg, M. C.*: Fracture of Glassy Polymers Within Sliding Contacts. Vol. 188, pp. 153–193.
Chen, P. see Jaffe, M.: Vol. 117, pp. 297–328.
Chern, C.-S. see Capek, I.: Vol. 155, pp. 101–166.
Chevolot, Y. see Mathieu, H. J.: Vol. 162, pp. 1–35.
Choe, E.-W. see Jaffe, M.: Vol. 117, pp. 297–328.
Chow, P. Y. and *Gan, L. M.*: Microemulsion Polymerizations and Reactions. Vol. 175, pp. 257–298.
Chow, T. S.: Glassy State Relaxation and Deformation in Polymers. Vol. 103, pp. 149–190.
Chujo, Y. see Uemura, T.: Vol. 167, pp. 81–106.
Chung, S.-J. see Lin, T.-C.: Vol. 161, pp. 157–193.
Chung, T.-S. see Jaffe, M.: Vol. 117, pp. 297–328.
Clarke, N.: Effect of Shear Flow on Polymer Blends. Vol. 183, pp. 127–173.
Coenjarts, C. see Li, M.: Vol. 190, pp. 183–226.
Cölfen, H. and *Antonietti, M.*: Field-Flow Fractionation Techniques for Polymer and Colloid Analysis. Vol. 150, pp. 67–187.
Colmenero, J. see Richter, D.: Vol. 174, pp. 1–221.
Comanita, B. see Roovers, J.: Vol. 142, pp. 179–228.
Comotti, A. see Sozzani, P.: Vol. 181, pp. 153–177.
Connell, J. W. see Hergenrother, P. M.: Vol. 117, pp. 67–110.

Corradini, P. see Auriemma, F.: Vol. 181, pp. 1–74.
Creton, C., Kramer, E. J., Brown, H. R. and *Hui, C.-Y.*: Adhesion and Fracture of Interfaces Between Immiscible Polymers: From the Molecular to the Continuum Scale. Vol. 156, pp. 53–135.
Criado-Sancho, M. see Jou, D.: Vol. 120, pp. 207–266.
Curro, J. G. see Schweizer, K. S.: Vol. 116, pp. 319–378.
Curtiss, C. F. and *Bird, R. B.*: Statistical Mechanics of Transport Phenomena: Polymeric Liquid Mixtures. Vol. 125, pp. 1–102.
Cussler, E. L., Wang, K. L. and *Burban, J. H.*: Hydrogels as Separation Agents. Vol. 110, pp. 67–80.
Czub, P. see Penczek, P.: Vol. 184, pp. 1–95.

Dalton, L.: Nonlinear Optical Polymeric Materials: From Chromophore Design to Commercial Applications. Vol. 158, pp. 1–86.
Dautzenberg, H. see Holm, C.: Vol. 166, pp. 113–171.
Davidson, J. M. see Prokop, A.: Vol. 160, pp. 119–174.
Den Decker, M. G. see Northolt, M. G.: Vol. 178, pp. 1–108.
Desai, S. M. and *Singh, R. P.*: Surface Modification of Polyethylene. Vol. 169, pp. 231–293.
DeSimone, J. M. see Canelas, D. A.: Vol. 133, pp. 103–140.
DeSimone, J. M. see Kennedy, K. A.: Vol. 175, pp. 329–346.
Dhal, P. K., Holmes-Farley, S. R., Huval, C. C. and *Jozefiak, T. H.*: Polymers as Drugs. Vol. 192, pp. 9–58.
DiMari, S. see Prokop, A.: Vol. 136, pp. 1–52.
Dimonie, M. V. see Hunkeler, D.: Vol. 112, pp. 115–134.
Dingenouts, N., Bolze, J., Pötschke, D. and *Ballauf, M.*: Analysis of Polymer Latexes by Small-Angle X-Ray Scattering. Vol. 144, pp. 1–48.
Dodd, L. R. and *Theodorou, D. N.*: Atomistic Monte Carlo Simulation and Continuum Mean Field Theory of the Structure and Equation of State Properties of Alkane and Polymer Melts. Vol. 116, pp. 249–282.
Doelker, E.: Cellulose Derivatives. Vol. 107, pp. 199–266.
Dolden, J. G.: Calculation of a Mesogenic Index with Emphasis Upon LC-Polyimides. Vol. 141, pp. 189–245.
Domb, A. J., Amselem, S., Shah, J. and *Maniar, M.*: Polyanhydrides: Synthesis and Characterization. Vol. 107, pp. 93–142.
Domb, A. J. see Kumar, M. N. V. R.: Vol. 160, pp. 45–118.
Doruker, P. see Baschnagel, J.: Vol. 152, pp. 41–156.
Dubois, P. see Mecerreyes, D.: Vol. 147, pp. 1–60.
Dubrovskii, S. A. see Kazanskii, K. S.: Vol. 104, pp. 97–134.
Dudowicz, J. see Freed, K. F.: Vol. 183, pp. 63–126.
Duncan, R., Ringsdorf, H. and *Satchi-Fainaro, R.*: Polymer Therapeutics: Polymers as Drugs, Drug and Protein Conjugates and Gene Delivery Systems: Past, Present and Future Opportunities. Vol. 192, pp. 1–8.
Dunkin, I. R. see Steinke, J.: Vol. 123, pp. 81–126.
Dunson, D. L. see McGrath, J. E.: Vol. 140, pp. 61–106.
Dziezok, P. see Rühe, J.: Vol. 165, pp. 79–150.

Eastmond, G. C.: Poly(e-caprolactone) Blends. Vol. 149, pp. 59–223.
Ebringerová, A., Hromádková, Z. and *Heinze, T.*: Hemicellulose. Vol. 186, pp. 1–67.
Economy, J. and *Goranov, K.*: Thermotropic Liquid Crystalline Polymers for High Performance Applications. Vol. 117, pp. 221–256.

Ediger, M. D. and *Adolf, D. B.*: Brownian Dynamics Simulations of Local Polymer Dynamics. Vol. 116, pp. 73–110.
Edlund, U. and *Albertsson, A.-C.*: Degradable Polymer Microspheres for Controlled Drug Delivery. Vol. 157, pp. 53–98.
Edwards, S. F. see Aharoni, S. M.: Vol. 118, pp. 1–231.
Eisenbach, C. D. see Bohrisch, J.: Vol. 165, pp. 1–41.
Endo, T. see Yagci, Y.: Vol. 127, pp. 59–86.
Engelhardt, H. and *Grosche, O.*: Capillary Electrophoresis in Polymer Analysis. Vol. 150, pp. 189–217.
Engelhardt, H. and *Martin, H.*: Characterization of Synthetic Polyelectrolytes by Capillary Electrophoretic Methods. Vol. 165, pp. 211–247.
Eriksson, P. see Jacobson, K.: Vol. 169, pp. 151–176.
Erman, B. see Bahar, I.: Vol. 116, pp. 145–206.
Eschner, M. see Spange, S.: Vol. 165, pp. 43–78.
Estel, K. see Spange, S.: Vol. 165, pp. 43–78.
Estevez, R. and *Van der Giessen, E.*: Modeling and Computational Analysis of Fracture of Glassy Polymers. Vol. 188, pp. 195–234.
Ewen, B. and *Richter, D.*: Neutron Spin Echo Investigations on the Segmental Dynamics of Polymers in Melts, Networks and Solutions. Vol. 134, pp. 1–130.
Ezquerra, T. A. see Baltá-Calleja, F. J.: Vol. 108, pp. 1–48.

Fatkullin, N. see Kimmich, R.: Vol. 170, pp. 1–113.
Faust, R. see Charleux, B.: Vol. 142, pp. 1–70.
Faust, R. see Kwon, Y.: Vol. 167, pp. 107–135.
Fekete, E. see Pukánszky, B.: Vol. 139, pp. 109–154.
Fendler, J. H.: Membrane-Mimetic Approach to Advanced Materials. Vol. 113, pp. 1–209.
Fetters, L. J. see Xu, Z.: Vol. 120, pp. 1–50.
Fontenot, K. see Schork, F. J.: Vol. 175, pp. 129–255.
Förster, S., Abetz, V. and *Müller, A. H. E.*: Polyelectrolyte Block Copolymer Micelles. Vol. 166, pp. 173–210.
Förster, S. and *Schmidt, M.*: Polyelectrolytes in Solution. Vol. 120, pp. 51–134.
Freed, K. F. and *Dudowicz, J.*: Influence of Monomer Molecular Structure on the Miscibility of Polymer Blends. Vol. 183, pp. 63–126.
Freire, J. J.: Conformational Properties of Branched Polymers: Theory and Simulations. Vol. 143, pp. 35–112.
Frenkel, D. see Hu, W.: Vol. 191, pp. 1–35.
Frenkel, S. Y. see Bronnikov, S. V.: Vol. 125, pp. 103–146.
Frick, B. see Baltá-Calleja, F. J.: Vol. 108, pp. 1–48.
Fridman, M. L.: see Terent'eva, J. P.: Vol. 101, pp. 29–64.
Fuchs, G. see Trimmel, G.: Vol. 176, pp. 43–87.
Fukui, K. see Otaigbe, J. U.: Vol. 154, pp. 1–86.
Funke, W.: Microgels-Intramolecularly Crosslinked Macromolecules with a Globular Structure. Vol. 136, pp. 137–232.
Furusho, Y. see Takata, T.: Vol. 171, pp. 1–75.
Furuya, H. see Abe, A.: Vol. 181, pp. 121–152.

Galina, H.: Mean-Field Kinetic Modeling of Polymerization: The Smoluchowski Coagulation Equation. Vol. 137, pp. 135–172.
Gan, L. M. see Chow, P. Y.: Vol. 175, pp. 257–298.
Ganesh, K. see Kishore, K.: Vol. 121, pp. 81–122.

Gaw, K. O. and *Kakimoto, M.*: Polyimide-Epoxy Composites. Vol. 140, pp. 107–136.
Geckeler, K. E. see Rivas, B.: Vol. 102, pp. 171–188.
Geckeler, K. E.: Soluble Polymer Supports for Liquid-Phase Synthesis. Vol. 121, pp. 31–80.
Gedde, U. W. and *Mattozzi, A.*: Polyethylene Morphology. Vol. 169, pp. 29–73.
Gehrke, S. H.: Synthesis, Equilibrium Swelling, Kinetics Permeability and Applications of Environmentally Responsive Gels. Vol. 110, pp. 81–144.
Geil, P. H., Yang, J., Williams, R. A., Petersen, K. L., Long, T.-C. and *Xu, P.*: Effect of Molecular Weight and Melt Time and Temperature on the Morphology of Poly(tetrafluorethylene). Vol. 180, pp. 89–159.
de Gennes, P.-G.: Flexible Polymers in Nanopores. Vol. 138, pp. 91–106.
Georgiou, S.: Laser Cleaning Methodologies of Polymer Substrates. Vol. 168, pp. 1–49.
Geuss, M. see Munz, M.: Vol. 164, pp. 87–210.
Giannelis, E. P., Krishnamoorti, R. and *Manias, E.*: Polymer-Silicate Nanocomposites: Model Systems for Confined Polymers and Polymer Brushes. Vol. 138, pp. 107–148.
Van der Giessen, E. see Estevez, R.: Vol. 188, pp. 195–234.
Godovsky, D. Y.: Device Applications of Polymer-Nanocomposites. Vol. 153, pp. 163–205.
Godovsky, D. Y.: Electron Behavior and Magnetic Properties Polymer-Nanocomposites. Vol. 119, pp. 79–122.
Gohy, J.-F.: Block Copolymer Micelles. Vol. 190, pp. 65–136.
González Arche, A. see Baltá-Calleja, F. J.: Vol. 108, pp. 1–48.
Goranov, K. see Economy, J.: Vol. 117, pp. 221–256.
Gramain, P. see Améduri, B.: Vol. 127, pp. 87–142.
Grein, C.: Toughness of Neat, Rubber Modified and Filled β-Nucleated Polypropylene: From Fundamentals to Applications. Vol. 188, pp. 43–104.
Grest, G. S.: Normal and Shear Forces Between Polymer Brushes. Vol. 138, pp. 149–184.
Grigorescu, G. and *Kulicke, W.-M.*: Prediction of Viscoelastic Properties and Shear Stability of Polymers in Solution. Vol. 152, p. 1–40.
Gröhn, F. see Rühe, J.: Vol. 165, pp. 79–150.
Grosberg, A. and *Nechaev, S.*: Polymer Topology. Vol. 106, pp. 1–30.
Grosche, O. see Engelhardt, H.: Vol. 150, pp. 189–217.
Grubbs, R., Risse, W. and *Novac, B.*: The Development of Well-defined Catalysts for Ring-Opening Olefin Metathesis. Vol. 102, pp. 47–72.
Gubler, U. and *Bosshard, C.*: Molecular Design for Third-Order Nonlinear Optics. Vol. 158, pp. 123–190.
Guida-Pietrasanta, F. and *Boutevin, B.*: Polysilalkylene or Silarylene Siloxanes Said Hybrid Silicones. Vol. 179, pp. 1–27.
van Gunsteren, W. F. see Gusev, A. A.: Vol. 116, pp. 207–248.
Gupta, B. and *Anjum, N.*: Plasma and Radiation-Induced Graft Modification of Polymers for Biomedical Applications. Vol. 162, pp. 37–63.
Gurtovenko, A. A. and *Blumen, A.*: Generalized Gaussian Structures: Models for Polymer Systems with Complex Topologies. Vol. 182, pp. 171–282.
Gusev, A. A., Müller-Plathe, F., van Gunsteren, W. F. and *Suter, U. W.*: Dynamics of Small Molecules in Bulk Polymers. Vol. 116, pp. 207–248.
Gusev, A. A. see Baschnagel, J.: Vol. 152, pp. 41–156.
Guillot, J. see Hunkeler, D.: Vol. 112, pp. 115–134.
Guyot, A. and *Tauer, K.*: Reactive Surfactants in Emulsion Polymerization. Vol. 111, pp. 43–66.

Hadjichristidis, N., Pispas, S., Pitsikalis, M., Iatrou, H. and *Vlahos, C.*: Asymmetric Star Polymers Synthesis and Properties. Vol. 142, pp. 71–128.

Hadjichristidis, N., Pitsikalis, M. and *Iatrou, H.*: Synthesis of Block Copolymers. Vol. 189, pp. 1–124.
Hadjichristidis, N. see *Xu, Z.*: Vol. 120, pp. 1–50.
Hadjichristidis, N. see *Pitsikalis, M.*: Vol. 135, pp. 1–138.
Hahn, O. see *Baschnagel, J.*: Vol. 152, pp. 41–156.
Hakkarainen, M.: Aliphatic Polyesters: Abiotic and Biotic Degradation and Degradation Products. Vol. 157, pp. 1–26.
Hakkarainen, M. and *Albertsson, A.-C.*: Environmental Degradation of Polyethylene. Vol. 169, pp. 177–199.
Halary, J. L. see *Monnerie, L.*: Vol. 187, pp. 35–213.
Halary, J. L. see *Monnerie, L.*: Vol. 187, pp. 215–364.
Hall, H. K. see *Penelle, J.*: Vol. 102, pp. 73–104.
Hamley, I. W.: Crystallization in Block Copolymers. Vol. 148, pp. 113–138.
Hammouda, B.: SANS from Homogeneous Polymer Mixtures: A Unified Overview. Vol. 106, pp. 87–134.
Han, M. J. and *Chang, J. Y.*: Polynucleotide Analogues. Vol. 153, pp. 1–36.
Harada, A.: Design and Construction of Supramolecular Architectures Consisting of Cyclodextrins and Polymers. Vol. 133, pp. 141–192.
Haralson, M. A. see *Prokop, A.*: Vol. 136, pp. 1–52.
Harding, S. E.: Analysis of Polysaccharides by Ultracentrifugation. Size, Conformation and Interactions in Solution. Vol. 186, pp. 211–254.
Hasegawa, N. see *Usuki, A.*: Vol. 179, pp. 135–195.
Hassan, C. M. and *Peppas, N. A.*: Structure and Applications of Poly(vinyl alcohol) Hydrogels Produced by Conventional Crosslinking or by Freezing/Thawing Methods. Vol. 153, pp. 37–65.
Hawker, C. J.: Dentritic and Hyperbranched Macromolecules Precisely Controlled Macromolecular Architectures. Vol. 147, pp. 113–160.
Hawker, C. J. see *Hedrick, J. L.*: Vol. 141, pp. 1–44.
He, G. S. see *Lin, T.-C.*: Vol. 161, pp. 157–193.
Hedrick, J. L., Carter, K. R., Labadie, J. W., Miller, R. D., Volksen, W., Hawker, C. J., Yoon, D. Y., Russell, T. P., McGrath, J. E. and *Briber, R. M.*: Nanoporous Polyimides. Vol. 141, pp. 1–44.
Hedrick, J. L., Labadie, J. W., Volksen, W. and *Hilborn, J. G.*: Nanoscopically Engineered Polyimides. Vol. 147, pp. 61–112.
Hedrick, J. L. see *Hergenrother, P. M.*: Vol. 117, pp. 67–110.
Hedrick, J. L. see *Kiefer, J.*: Vol. 147, pp. 161–247.
Hedrick, J. L. see *McGrath, J. E.*: Vol. 140, pp. 61–106.
Heine, D. R., Grest, G. S. and *Curro, J. G.*: Structure of Polymer Melts and Blends: Comparison of Integral Equation theory and Computer Sumulation. Vol. 173, pp. 209–249.
Heinrich, G. and *Klüppel, M.*: Recent Advances in the Theory of Filler Networking in Elastomers. Vol. 160, pp. 1–44.
Heinze, T. see *Ebringerová, A.*: Vol. 186, pp. 1–67.
Heinze, T. see *El Seoud, O. A.*: Vol. 186, pp. 103–149.
Heller, J.: Poly (Ortho Esters). Vol. 107, pp. 41–92.
Helm, C. A. see *Möhwald, H.*: Vol. 165, pp. 151–175.
Hemielec, A. A. see *Hunkeler, D.*: Vol. 112, pp. 115–134.
Hergenrother, P. M., Connell, J. W., Labadie, J. W. and *Hedrick, J. L.*: Poly(arylene ether)s Containing Heterocyclic Units. Vol. 117, pp. 67–110.
Hernández-Barajas, J. see *Wandrey, C.*: Vol. 145, pp. 123–182.
Hervet, H. see *Léger, L.*: Vol. 138, pp. 185–226.
Hiejima, T. see *Abe, A.*: Vol. 181, pp. 121–152.

Hikosaka, M., Watanabe, K., Okada, K. and *Yamazaki, S.*: Topological Mechanism of Polymer Nucleation and Growth – The Role of Chain Sliding Diffusion and Entanglement. Vol. 191, pp. 137–186.
Hilborn, J. G. see Hedrick, J. L.: Vol. 147, pp. 61–112.
Hilborn, J. G. see Kiefer, J.: Vol. 147, pp. 161–247.
Hillborg, H. see Vancso, G. J.: Vol. 182, pp. 55–129.
Hillmyer, M. A.: Nanoporous Materials from Block Copolymer Precursors. Vol. 190, pp. 137–181.
Hiramatsu, N. see Matsushige, M.: Vol. 125, pp. 147–186.
Hirasa, O. see Suzuki, M.: Vol. 110, pp. 241–262.
Hirotsu, S.: Coexistence of Phases and the Nature of First-Order Transition in Poly-N-isopropylacrylamide Gels. Vol. 110, pp. 1–26.
Höcker, H. see Klee, D.: Vol. 149, pp. 1–57.
Holm, C. see Arnold, A.: Vol. 185, pp. 59–109.
Holm, C., Hofmann, T., Joanny, J. F., Kremer, K., Netz, R. R., Reineker, P., Seidel, C., Vilgis, T. A. and *Winkler, R. G.*: Polyelectrolyte Theory. Vol. 166, pp. 67–111.
Holm, C., Rehahn, M., Oppermann, W. and *Ballauff, M.*: Stiff-Chain Polyelectrolytes. Vol. 166, pp. 1–27.
Holmes-Farley, S. R. see Dhal, P. K.: Vol. 192, pp. 9–58.
Hornsby, P.: Rheology, Compounding and Processing of Filled Thermoplastics. Vol. 139, pp. 155–216.
Houbenov, N. see Rühe, J.: Vol. 165, pp. 79–150.
Hromádková, Z. see Ebringerová, A.: Vol. 186, pp. 1–67.
Hu, W. and *Frenkel, D.*: Polymer Crystallization Driven by Anisotropic Interactions. Vol. 191, pp. 1–35.
Huber, K. see Volk, N.: Vol. 166, pp. 29–65.
Hugenberg, N. see Rühe, J.: Vol. 165, pp. 79–150.
Hui, C.-Y. see Creton, C.: Vol. 156, pp. 53–135.
Hult, A., Johansson, M. and *Malmström, E.*: Hyperbranched Polymers. Vol. 143, pp. 1–34.
Hünenberger, P. H.: Thermostat Algorithms for Molecular-Dynamics Simulations. Vol. 173, pp. 105–147.
Hunkeler, D., Candau, F., Pichot, C., Hemielec, A. E., Xie, T. Y., Barton, J., Vaskova, V., Guillot, J., Dimonie, M. V. and *Reichert, K. H.*: Heterophase Polymerization: A Physical and Kinetic Comparision and Categorization. Vol. 112, pp. 115–134.
Hunkeler, D. see Macko, T.: Vol. 163, pp. 61–136.
Hunkeler, D. see Prokop, A.: Vol. 136, pp. 1–52; 53–74.
Hunkeler, D. see Wandrey, C.: Vol. 145, pp. 123–182.
Huval, C. C. see Dhal, P. K.: Vol. 192, pp. 9–58.

Iatrou, H. see Hadjichristidis, N.: Vol. 142, pp. 71–128.
Iatrou, H. see Hadjichristidis, N.: Vol. 189, pp. 1–124.
Ichikawa, T. see Yoshida, H.: Vol. 105, pp. 3–36.
Ihara, E. see Yasuda, H.: Vol. 133, pp. 53–102.
Ikada, Y. see Uyama, Y.: Vol. 137, pp. 1–40.
Ikehara, T. see Jinnuai, H.: Vol. 170, pp. 115–167.
Ilavsky, M.: Effect on Phase Transition on Swelling and Mechanical Behavior of Synthetic Hydrogels. Vol. 109, pp. 173–206.
Imai, M. see Kaji, K.: Vol. 191, pp. 187–240.
Imai, Y.: Rapid Synthesis of Polyimides from Nylon-Salt Monomers. Vol. 140, pp. 1–23.
Inomata, H. see Saito, S.: Vol. 106, pp. 207–232.

Inoue, S. see Sugimoto, H.: Vol. 146, pp. 39–120.
Irie, M.: Stimuli-Responsive Poly(N-isopropylacrylamide), Photo- and Chemical-Induced Phase Transitions. Vol. 110, pp. 49–66.
Ise, N. see Matsuoka, H.: Vol. 114, pp. 187–232.
Ishikawa, T.: Advances in Inorganic Fibers. Vol. 178, pp. 109–144.
Ito, H.: Chemical Amplification Resists for Microlithography. Vol. 172, pp. 37–245.
Ito, K. and *Kawaguchi, S.*: Poly(macronomers), Homo- and Copolymerization. Vol. 142, pp. 129–178.
Ito, K. see Kawaguchi, S.: Vol. 175, pp. 299–328.
Ito, S. and *Aoki, H.*: Nano-Imaging of Polymers by Optical Microscopy. Vol. 182, pp. 131–170.
Ito, Y. see Suginome, M.: Vol. 171, pp. 77–136.
Ivanov, A. E. see Zubov, V. P.: Vol. 104, pp. 135–176.

Jacob, S. and *Kennedy, J.*: Synthesis, Characterization and Properties of OCTA-ARM Polyisobutylene-Based Star Polymers. Vol. 146, pp. 1–38.
Jacobson, K., Eriksson, P., Reitberger, T. and *Stenberg, B.*: Chemiluminescence as a Tool for Polyolefin. Vol. 169, pp. 151–176.
Jaeger, W. see Bohrisch, J.: Vol. 165, pp. 1–41.
Jaffe, M., Chen, P., Choe, E.-W., Chung, T.-S. and *Makhija, S.*: High Performance Polymer Blends. Vol. 117, pp. 297–328.
Jancar, J.: Structure-Property Relationships in Thermoplastic Matrices. Vol. 139, pp. 1–66.
Jen, A. K.-Y. see Kajzar, F.: Vol. 161, pp. 1–85.
Jerome, R. see Mecerreyes, D.: Vol. 147, pp. 1–60.
de Jeu, W. H. see Li, L.: Vol. 181, pp. 75–120.
Jiang, M., Li, M., Xiang, M. and *Zhou, H.*: Interpolymer Complexation and Miscibility and Enhancement by Hydrogen Bonding. Vol. 146, pp. 121–194.
Jin, J. see Shim, H.-K.: Vol. 158, pp. 191–241.
Jinnai, H., Nishikawa, Y., Ikehara, T. and *Nishi, T.*: Emerging Technologies for the 3D Analysis of Polymer Structures. Vol. 170, pp. 115–167.
Jo, W. H. and *Yang, J. S.*: Molecular Simulation Approaches for Multiphase Polymer Systems. Vol. 156, pp. 1–52.
Joanny, J.-F. see Holm, C.: Vol. 166, pp. 67–111.
Joanny, J.-F. see Thünemann, A. F.: Vol. 166, pp. 113–171.
Johannsmann, D. see Rühe, J.: Vol. 165, pp. 79–150.
Johansson, M. see Hult, A.: Vol. 143, pp. 1–34.
Joos-Müller, B. see Funke, W.: Vol. 136, pp. 137–232.
Jou, D., Casas-Vazquez, J. and *Criado-Sancho, M.*: Thermodynamics of Polymer Solutions under Flow: Phase Separation and Polymer Degradation. Vol. 120, pp. 207–266.
Jozefiak, T. H. see Dhal, P. K.: Vol. 192, pp. 9–58.

Kaetsu, I.: Radiation Synthesis of Polymeric Materials for Biomedical and Biochemical Applications. Vol. 105, pp. 81–98.
Kaji, K., Nishida, K., Kanaya, T., Matsuba, G., Konishi, T. and *Imai, M.*: Spinodal Crystallization of Polymers: Crystallization from the Unstable Melt. Vol. 191, pp. 187–240.
Kaji, K. see Kanaya, T.: Vol. 154, pp. 87–141.
Kajzar, F., Lee, K.-S. and *Jen, A. K.-Y.*: Polymeric Materials and their Orientation Techniques for Second-Order Nonlinear Optics. Vol. 161, pp. 1–85.
Kakimoto, M. see Gaw, K. O.: Vol. 140, pp. 107–136.
Kaminski, W. and *Arndt, M.*: Metallocenes for Polymer Catalysis. Vol. 127, pp. 143–187.

Kammer, H. W., Kressler, H. and *Kummerloewe, C.*: Phase Behavior of Polymer Blends – Effects of Thermodynamics and Rheology. Vol. 106, pp. 31–86.
Kanaya, T. and *Kaji, K.*: Dynamcis in the Glassy State and Near the Glass Transition of Amorphous Polymers as Studied by Neutron Scattering. Vol. 154, pp. 87–141.
Kanaya, T. see Kaji, K.: Vol. 191, pp. 187–240.
Kandyrin, L. B. and *Kuleznev, V. N.*: The Dependence of Viscosity on the Composition of Concentrated Dispersions and the Free Volume Concept of Disperse Systems. Vol. 103, pp. 103–148.
Kaneko, M. see Ramaraj, R.: Vol. 123, pp. 215–242.
Kang, E. T., Neoh, K. G. and *Tan, K. L.*: X-Ray Photoelectron Spectroscopic Studies of Electroactive Polymers. Vol. 106, pp. 135–190.
Karlsson, S. see Söderqvist Lindblad, M.: Vol. 157, pp. 139–161.
Karlsson, S.: Recycled Polyolefins. Material Properties and Means for Quality Determination. Vol. 169, pp. 201–229.
Kato, K. see Uyama, Y.: Vol. 137, pp. 1–40.
Kato, M. see Usuki, A.: Vol. 179, pp. 135–195.
Kausch, H.-H. and *Michler, G. H.*: The Effect of Time on Crazing and Fracture. Vol. 187, pp. 1–33.
Kausch, H.-H. see Monnerie, L. Vol. 187, pp. 215–364.
Kautek, W. see Krüger, J.: Vol. 168, pp. 247–290.
Kawaguchi, S. see Ito, K.: Vol. 142, pp. 129–178.
Kawaguchi, S. and *Ito, K.*: Dispersion Polymerization. Vol. 175, pp. 299–328.
Kawata, S. see Sun, H.-B.: Vol. 170, pp. 169–273.
Kazanskii, K. S. and *Dubrovskii, S. A.*: Chemistry and Physics of Agricultural Hydrogels. Vol. 104, pp. 97–134.
Kennedy, J. P. see Jacob, S.: Vol. 146, pp. 1–38.
Kennedy, J. P. see Majoros, I.: Vol. 112, pp. 1–113.
Kennedy, K. A., Roberts, G. W. and *DeSimone, J. M.*: Heterogeneous Polymerization of Fluoroolefins in Supercritical Carbon Dioxide. Vol. 175, pp. 329–346.
Khokhlov, A., Starodybtzev, S. and *Vasilevskaya, V.*: Conformational Transitions of Polymer Gels: Theory and Experiment. Vol. 109, pp. 121–172.
Kiefer, J., Hedrick, J. L. and *Hiborn, J. G.*: Macroporous Thermosets by Chemically Induced Phase Separation. Vol. 147, pp. 161–247.
Kihara, N. see Takata, T.: Vol. 171, pp. 1–75.
Kilian, H. G. and *Pieper, T.*: Packing of Chain Segments. A Method for Describing X-Ray Patterns of Crystalline, Liquid Crystalline and Non-Crystalline Polymers. Vol. 108, pp. 49–90.
Kim, J. see Quirk, R. P.: Vol. 153, pp. 67–162.
Kim, K.-S. see Lin, T.-C.: Vol. 161, pp. 157–193.
Kimmich, R. and *Fatkullin, N.*: Polymer Chain Dynamics and NMR. Vol. 170, pp. 1–113.
Kippelen, B. and *Peyghambarian, N.*: Photorefractive Polymers and their Applications. Vol. 161, pp. 87–156.
Kirchhoff, R. A. and *Bruza, K. J.*: Polymers from Benzocyclobutenes. Vol. 117, pp. 1–66.
Kishore, K. and *Ganesh, K.*: Polymers Containing Disulfide, Tetrasulfide, Diselenide and Ditelluride Linkages in the Main Chain. Vol. 121, pp. 81–122.
Kitamaru, R.: Phase Structure of Polyethylene and Other Crystalline Polymers by Solid-State 13C/MNR. Vol. 137, pp. 41–102.
Klapper, M. see Rusanov, A. L.: Vol. 179, pp. 83–134.
Klee, D. and *Höcker, H.*: Polymers for Biomedical Applications: Improvement of the Interface Compatibility. Vol. 149, pp. 1–57.

Klemm, E., Pautzsch, T. and *Blankenburg, L.*: Organometallic PAEs. Vol. 177, pp. 53–90.
Klier, J. see Scranton, A. B.: Vol. 122, pp. 1–54.
v. Klitzing, R. and *Tieke, B.*: Polyelectrolyte Membranes. Vol. 165, pp. 177–210.
Kloeckner, J. see Wagner, E.: Vol. 192, pp. 135–173.
Klüppel, M.: The Role of Disorder in Filler Reinforcement of Elastomers on Various Length Scales. Vol. 164, pp. 1–86.
Klüppel, M. see Heinrich, G.: Vol. 160, pp. 1–44.
Knuuttila, H., Lehtinen, A. and *Nummila-Pakarinen, A.*: Advanced Polyethylene Technologies – Controlled Material Properties. Vol. 169, pp. 13–27.
Kobayashi, S., Shoda, S. and *Uyama, H.*: Enzymatic Polymerization and Oligomerization. Vol. 121, pp. 1–30.
Kobayashi, T. see Abe, A.: Vol. 181, pp. 121–152.
Köhler, W. and *Schäfer, R.*: Polymer Analysis by Thermal-Diffusion Forced Rayleigh Scattering. Vol. 151, pp. 1–59.
Koenig, J. L. see Bhargava, R.: Vol. 163, pp. 137–191.
Koenig, J. L. see Andreis, M.: Vol. 124, pp. 191–238.
Koike, T.: Viscoelastic Behavior of Epoxy Resins Before Crosslinking. Vol. 148, pp. 139–188.
Kokko, E. see Löfgren, B.: Vol. 169, pp. 1–12.
Kokufuta, E.: Novel Applications for Stimulus-Sensitive Polymer Gels in the Preparation of Functional Immobilized Biocatalysts. Vol. 110, pp. 157–178.
Konishi, T. see Kaji, K.: Vol. 191, pp. 187–240.
Konno, M. see Saito, S.: Vol. 109, pp. 207–232.
Konradi, R. see Rühe, J.: Vol. 165, pp. 79–150.
Kopecek, J. see Putnam, D.: Vol. 122, pp. 55–124.
Koßmehl, G. see Schopf, G.: Vol. 129, pp. 1–145.
Kostoglodov, P. V. see Rusanov, A. L.: Vol. 179, pp. 83–134.
Kozlov, E. see Prokop, A.: Vol. 160, pp. 119–174.
Kramer, E. J. see Creton, C.: Vol. 156, pp. 53–135.
Kremer, K. see Baschnagel, J.: Vol. 152, pp. 41–156.
Kremer, K. see Holm, C.: Vol. 166, pp. 67–111.
Kressler, J. see Kammer, H. W.: Vol. 106, pp. 31–86.
Kricheldorf, H. R.: Liquid-Crystalline Polyimides. Vol. 141, pp. 83–188.
Krishnamoorti, R. see Giannelis, E. P.: Vol. 138, pp. 107–148.
Krüger, J. and *Kautek, W.*: Ultrashort Pulse Laser Interaction with Dielectrics and Polymers, Vol. 168, pp. 247–290.
Kuchanov, S. I.: Modern Aspects of Quantitative Theory of Free-Radical Copolymerization. Vol. 103, pp. 1–102.
Kuchanov, S. I.: Principles of Quantitive Description of Chemical Structure of Synthetic Polymers. Vol. 152, pp. 157–202.
Kudaibergennow, S. E.: Recent Advances in Studying of Synthetic Polyampholytes in Solutions. Vol. 144, pp. 115–198.
Kuleznev, V. N. see Kandyrin, L. B.: Vol. 103, pp. 103–148.
Kulichkhin, S. G. see Malkin, A. Y.: Vol. 101, pp. 217–258.
Kulicke, W.-M. see Grigorescu, G.: Vol. 152, pp. 1–40.
Kumar, M. N. V. R., Kumar, N., Domb, A. J. and *Arora, M.*: Pharmaceutical Polymeric Controlled Drug Delivery Systems. Vol. 160, pp. 45–118.
Kumar, N. see Kumar, M. N. V. R.: Vol. 160, pp. 45–118.
Kummerloewe, C. see Kammer, H. W.: Vol. 106, pp. 31–86.
Kuznetsova, N. P. see Samsonov, G. V.: Vol. 104, pp. 1–50.

Kwon, Y. and *Faust, R.*: Synthesis of Polyisobutylene-Based Block Copolymers with Precisely Controlled Architecture by Living Cationic Polymerization. Vol. 167, pp. 107–135.

Labadie, J. W. see Hergenrother, P. M.: Vol. 117, pp. 67–110.
Labadie, J. W. see Hedrick, J. L.: Vol. 141, pp. 1–44.
Labadie, J. W. see Hedrick, J. L.: Vol. 147, pp. 61–112.
Lamparski, H. G. see O'Brien, D. F.: Vol. 126, pp. 53–84.
Laschewsky, A.: Molecular Concepts, Self-Organisation and Properties of Polysoaps. Vol. 124, pp. 1–86.
Laso, M. see Leontidis, E.: Vol. 116, pp. 283–318.
Laupêtre, F. see Monnerie, L.: Vol. 187, pp. 35–213.
Lazár, M. and *Rychl, R.*: Oxidation of Hydrocarbon Polymers. Vol. 102, pp. 189–222.
Lechowicz, J. see Galina, H.: Vol. 137, pp. 135–172.
Léger, L., Raphaël, E. and *Hervet, H.*: Surface-Anchored Polymer Chains: Their Role in Adhesion and Friction. Vol. 138, pp. 185–226.
Lenz, R. W.: Biodegradable Polymers. Vol. 107, pp. 1–40.
Leontidis, E., de Pablo, J. J., Laso, M. and *Suter, U. W.*: A Critical Evaluation of Novel Algorithms for the Off-Lattice Monte Carlo Simulation of Condensed Polymer Phases. Vol. 116, pp. 283–318.
Lee, B. see Quirk, R. P.: Vol. 153, pp. 67–162.
Lee, K.-S. see Kajzar, F.: Vol. 161, pp. 1–85.
Lee, Y. see Quirk, R. P.: Vol. 153, pp. 67–162.
Lehtinen, A. see Knuuttila, H.: Vol. 169, pp. 13–27.
Leónard, D. see Mathieu, H. J.: Vol. 162, pp. 1–35.
Lesec, J. see Viovy, J.-L.: Vol. 114, pp. 1–42.
Levesque, D. see Weis, J.-J.: Vol. 185, pp. 163–225.
Li, L. and *de Jeu, W. H.*: Flow-induced mesophases in crystallizable polymers. Vol. 181, pp. 75–120.
Li, L. see Chan, C.-M.: Vol. 188, pp. 1–41.
Li, M., Coenjarts, C. and *Ober, C. K.*: Patternable Block Copolymers. Vol. 190, pp. 183–226.
Li, M. see Jiang, M.: Vol. 146, pp. 121–194.
Liang, G. L. see Sumpter, B. G.: Vol. 116, pp. 27–72.
Lienert, K.-W.: Poly(ester-imide)s for Industrial Use. Vol. 141, pp. 45–82.
Likhatchev, D. see Rusanov, A. L.: Vol. 179, pp. 83–134.
Lin, J. and *Sherrington, D. C.*: Recent Developments in the Synthesis, Thermostability and Liquid Crystal Properties of Aromatic Polyamides. Vol. 111, pp. 177–220.
Lin, T.-C., Chung, S.-J., Kim, K.-S., Wang, X., He, G. S., Swiatkiewicz, J., Pudavar, H. E. and *Prasad, P. N.*: Organics and Polymers with High Two-Photon Activities and their Applications. Vol. 161, pp. 157–193.
Linse, P.: Simulation of Charged Colloids in Solution. Vol. 185, pp. 111–162.
Lippert, T.: Laser Application of Polymers. Vol. 168, pp. 51–246.
Liu, Y. see Söderqvist Lindblad, M.: Vol. 157, pp. 139–161.
Long, T.-C. see Geil, P. H.: Vol. 180, pp. 89–159.
López Cabarcos, E. see Baltá-Calleja, F. J.: Vol. 108, pp. 1–48.
Lotz, B.: Analysis and Observation of Polymer Crystal Structures at the Individual Stem Level. Vol. 180, pp. 17–44.
Löfgren, B., Kokko, E. and *Seppälä, J.*: Specific Structures Enabled by Metallocene Catalysis in Polyethenes. Vol. 169, pp. 1–12.
Löwen, H. see Thünemann, A. F.: Vol. 166, pp. 113–171.
Luo, Y. see Schork, F. J.: Vol. 175, pp. 129–255.

Macko, T. and *Hunkeler, D.*: Liquid Chromatography under Critical and Limiting Conditions: A Survey of Experimental Systems for Synthetic Polymers. Vol. 163, pp. 61–136.

Majoros, I., Nagy, A. and *Kennedy, J. P.*: Conventional and Living Carbocationic Polymerizations United. I. A Comprehensive Model and New Diagnostic Method to Probe the Mechanism of Homopolymerizations. Vol. 112, pp. 1–113.

Makhija, S. see Jaffe, M.: Vol. 117, pp. 297–328.

Malmström, E. see Hult, A.: Vol. 143, pp. 1–34.

Malkin, A. Y. and *Kulichkhin, S. G.*: Rheokinetics of Curing. Vol. 101, pp. 217–258.

Maniar, M. see Domb, A. J.: Vol. 107, pp. 93–142.

Manias, E. see Giannelis, E. P.: Vol. 138, pp. 107–148.

Martin, H. see Engelhardt, H.: Vol. 165, pp. 211–247.

Marty, J. D. and *Mauzac, M.*: Molecular Imprinting: State of the Art and Perspectives. Vol. 172, pp. 1–35.

Mashima, K., Nakayama, Y. and *Nakamura, A.*: Recent Trends in Polymerization of a-Olefins Catalyzed by Organometallic Complexes of Early Transition Metals. Vol. 133, pp. 1–52.

Mathew, D. see Reghunadhan Nair, C. P.: Vol. 155, pp. 1–99.

Mathieu, H. J., Chevolot, Y., Ruiz-Taylor, L. and *Leónard, D.*: Engineering and Characterization of Polymer Surfaces for Biomedical Applications. Vol. 162, pp. 1–35.

Matsuba, G. see Kaji, K.: Vol. 191, pp. 187–240.

Matsumoto, A.: Free-Radical Crosslinking Polymerization and Copolymerization of Multivinyl Compounds. Vol. 123, pp. 41–80.

Matsumoto, A. see Otsu, T.: Vol. 136, pp. 75–138.

Matsuoka, H. and *Ise, N.*: Small-Angle and Ultra-Small Angle Scattering Study of the Ordered Structure in Polyelectrolyte Solutions and Colloidal Dispersions. Vol. 114, pp. 187–232.

Matsushige, K., Hiramatsu, N. and *Okabe, H.*: Ultrasonic Spectroscopy for Polymeric Materials. Vol. 125, pp. 147–186.

Mattice, W. L. see Rehahn, M.: Vol. 131/132, pp. 1–475.

Mattice, W. L. see Baschnagel, J.: Vol. 152, pp. 41–156.

Mattozzi, A. see Gedde, U. W.: Vol. 169, pp. 29–73.

Mauzac, M. see Marty, J. D.: Vol. 172, pp. 1–35.

Mays, W. see Xu, Z.: Vol. 120, pp. 1–50.

Mays, J. W. see Pitsikalis, M.: Vol. 135, pp. 1–138.

McGrath, J. E. see Hedrick, J. L.: Vol. 141, pp. 1–44.

McGrath, J. E., Dunson, D. L. and *Hedrick, J. L.*: Synthesis and Characterization of Segmented Polyimide-Polyorganosiloxane Copolymers. Vol. 140, pp. 61–106.

McLeish, T. C. B. and *Milner, S. T.*: Entangled Dynamics and Melt Flow of Branched Polymers. Vol. 143, pp. 195–256.

Mecerreyes, D., Dubois, P. and *Jerome, R.*: Novel Macromolecular Architectures Based on Aliphatic Polyesters: Relevance of the Coordination-Insertion Ring-Opening Polymerization. Vol. 147, pp. 1–60.

Mecham, S. J. see McGrath, J. E.: Vol. 140, pp. 61–106.

Meille, S. V. see Allegra, G.: Vol. 191, pp. 87–135.

Menzel, H. see Möhwald, H.: Vol. 165, pp. 151–175.

Meyer, T. see Spange, S.: Vol. 165, pp. 43–78.

Michler, G. H. see Kausch, H.-H.: Vol. 187, pp. 1–33.

Mikos, A. G. see Thomson, R. C.: Vol. 122, pp. 245–274.

Milner, S. T. see McLeish, T. C. B.: Vol. 143, pp. 195–256.

Mison, P. and *Sillion, B.*: Thermosetting Oligomers Containing Maleimides and Nadiimides End-Groups. Vol. 140, pp. 137–180.

Miyasaka, K.: PVA-Iodine Complexes: Formation, Structure and Properties. Vol. 108, pp. 91–130.
Miller, R. D. see Hedrick, J. L.: Vol. 141, pp. 1–44.
Minko, S. see Rühe, J.: Vol. 165, pp. 79–150.
Möhwald, H., Menzel, H., Helm, C. A. and *Stamm, M.*: Lipid and Polyampholyte Monolayers to Study Polyelectrolyte Interactions and Structure at Interfaces. Vol. 165, pp. 151–175.
Monkenbusch, M. see Richter, D.: Vol. 174, pp. 1–221.
Monnerie, L., Halary, J. L. and *Kausch, H.-H.*: Deformation, Yield and Fracture of Amorphous Polymers: Relation to the Secondary Transitions. Vol. 187, pp. 215–364.
Monnerie, L., Lauprêtre, F. and *Halary, J. L.*: Investigation of Solid-State Transitions in Linear and Crosslinked Amorphous Polymers. Vol. 187, pp. 35–213.
Monnerie, L. see Bahar, I.: Vol. 116, pp. 145–206.
Moore, J. S. see Ray, C. R.: Vol. 177, pp. 99–149.
Mori, H. see Bohrisch, J.: Vol. 165, pp. 1–41.
Morishima, Y.: Photoinduced Electron Transfer in Amphiphilic Polyelectrolyte Systems. Vol. 104, pp. 51–96.
Morton, M. see Quirk, R. P.: Vol. 153, pp. 67–162.
Motornov, M. see Rühe, J.: Vol. 165, pp. 79–150.
Mours, M. see Winter, H. H.: Vol. 134, pp. 165–234.
Müllen, K. see Scherf, U.: Vol. 123, pp. 1–40.
Müller, A. H. E. see Bohrisch, J.: Vol. 165, pp. 1–41.
Müller, A. H. E. see Förster, S.: Vol. 166, pp. 173–210.
Müller, A. J., Balsamo, V. and *Arnal, M. L.*: Nucleation and Crystallization in Diblock and Triblock Copolymers. Vol. 190, pp. 1–63.
Müller, M. and *Schmid, F.*: Incorporating Fluctuations and Dynamics in Self-Consistent Field Theories for Polymer Blends. Vol. 185, pp. 1–58.
Müller, M. see Thünemann, A. F.: Vol. 166, pp. 113–171.
Müller-Plathe, F. see Gusev, A. A.: Vol. 116, pp. 207–248.
Müller-Plathe, F. see Baschnagel, J.: Vol. 152, p. 41–156.
Mukerherjee, A. see Biswas, M.: Vol. 115, pp. 89–124.
Munz, M., Cappella, B., Sturm, H., Geuss, M. and *Schulz, E.*: Materials Contrasts and Nanolithography Techniques in Scanning Force Microscopy (SFM) and their Application to Polymers and Polymer Composites. Vol. 164, pp. 87–210.
Murat, M. see Baschnagel, J.: Vol. 152, p. 41–156.
Muthukumar, M.: Modeling Polymer Crystallization. Vol. 191, pp. 241–274.
Muzzarelli, C. see Muzzarelli, R. A. A.: Vol. 186, pp. 151–209.
Muzzarelli, R. A. A. and *Muzzarelli, C.*: Chitosan Chemistry: Relevance to the Biomedical Sciences. Vol. 186, pp. 151–209.
Mylnikov, V.: Photoconducting Polymers. Vol. 115, pp. 1–88.

Nagy, A. see Majoros, I.: Vol. 112, pp. 1–11.
Naka, K. see Uemura, T.: Vol. 167, pp. 81–106.
Nakamura, A. see Mashima, K.: Vol. 133, pp. 1–52.
Nakayama, Y. see Mashima, K.: Vol. 133, pp. 1–52.
Narasinham, B. and *Peppas, N. A.*: The Physics of Polymer Dissolution: Modeling Approaches and Experimental Behavior. Vol. 128, pp. 157–208.
Nechaev, S. see Grosberg, A.: Vol. 106, pp. 1–30.
Neoh, K. G. see Kang, E. T.: Vol. 106, pp. 135–190.
Netz, R. R. see Holm, C.: Vol. 166, pp. 67–111.

Netz, R. R. see Rühe, J.: Vol. 165, pp. 79–150.
Newman, S. M. see Anseth, K. S.: Vol. 122, pp. 177–218.
Nijenhuis, K. te: Thermoreversible Networks. Vol. 130, pp. 1–252.
Ninan, K. N. see Reghunadhan Nair, C. P.: Vol. 155, pp. 1–99.
Nishi, T. see Jinnai, H.: Vol. 170, pp. 115–167.
Nishida, K. see Kaji, K.: Vol. 191, pp. 187–240.
Nishikawa, Y. see Jinnai, H.: Vol. 170, pp. 115–167.
Noid, D. W. see Otaigbe, J. U.: Vol. 154, pp. 1–86.
Noid, D. W. see Sumpter, B. G.: Vol. 116, pp. 27–72.
Nomura, M., Tobita, H. and *Suzuki, K.*: Emulsion Polymerization: Kinetic and Mechanistic Aspects. Vol. 175, pp. 1–128.
Northolt, M. G., Picken, S. J., Den Decker, M. G., Baltussen, J. J. M. and *Schlatmann, R.*: The Tensile Strength of Polymer Fibres. Vol. 178, pp. 1–108.
Novac, B. see Grubbs, R.: Vol. 102, pp. 47–72.
Novikov, V. V. see Privalko, V. P.: Vol. 119, pp. 31–78.
Nummila-Pakarinen, A. see Knuuttila, H.: Vol. 169, pp. 13–27.

Ober, C. K. see Li, M.: Vol. 190, pp. 183–226.
O'Brien, D. F., Armitage, B. A., Bennett, D. E. and *Lamparski, H. G.*: Polymerization and Domain Formation in Lipid Assemblies. Vol. 126, pp. 53–84.
Ogasawara, M.: Application of Pulse Radiolysis to the Study of Polymers and Polymerizations. Vol.105, pp. 37–80.
Okabe, H. see Matsushige, K.: Vol. 125, pp. 147–186.
Okada, M.: Ring-Opening Polymerization of Bicyclic and Spiro Compounds. Reactivities and Polymerization Mechanisms. Vol. 102, pp. 1–46.
Okada, K. see Hikosaka, M.: Vol. 191, pp. 137–186.
Okano, T.: Molecular Design of Temperature-Responsive Polymers as Intelligent Materials. Vol. 110, pp. 179–198.
Okay, O. see Funke, W.: Vol. 136, pp. 137–232.
Onuki, A.: Theory of Phase Transition in Polymer Gels. Vol. 109, pp. 63–120.
Oppermann, W. see Holm, C.: Vol. 166, pp. 1–27.
Oppermann, W. see Volk, N.: Vol. 166, pp. 29–65.
Osad'ko, I. S.: Selective Spectroscopy of Chromophore Doped Polymers and Glasses. Vol. 114, pp. 123–186.
Osakada, K. and *Takeuchi, D.*: Coordination Polymerization of Dienes, Allenes, and Methylenecycloalkanes. Vol. 171, pp. 137–194.
Otaigbe, J. U., Barnes, M. D., Fukui, K., Sumpter, B. G. and *Noid, D. W.*: Generation, Characterization, and Modeling of Polymer Micro- and Nano-Particles. Vol. 154, pp. 1–86.
Otsu, T. and *Matsumoto, A.*: Controlled Synthesis of Polymers Using the Iniferter Technique: Developments in Living Radical Polymerization. Vol. 136, pp. 75–138.

de Pablo, J. J. see Leontidis, E.: Vol. 116, pp. 283–318.
Padias, A. B. see Penelle, J.: Vol. 102, pp. 73–104.
Pascault, J.-P. see Williams, R. J. J.: Vol. 128, pp. 95–156.
Pasch, H.: Analysis of Complex Polymers by Interaction Chromatography. Vol. 128, pp. 1–46.
Pasch, H.: Hyphenated Techniques in Liquid Chromatography of Polymers. Vol. 150, pp. 1–66.
Pasut, G. and *Veronese, F. M.*: PEGylation of Proteins as Tailored Chemistry for Optimized Bioconjugates. Vol. 192, pp. 95–134.
Paul, W. see Baschnagel, J.: Vol. 152, pp. 41–156.

Paulsen, S. B. and *Barsett, H.*: Bioactive Pectic Polysaccharides. Vol. 186, pp. 69–101.
Pautzsch, T. see Klemm, E.: Vol. 177, pp. 53–90.
Penczek, P., Czub, P. and *Pielichowski, J.*: Unsaturated Polyester Resins: Chemistry and Technology. Vol. 184, pp. 1–95.
Penczek, P. see Batog, A. E.: Vol. 144, pp. 49–114.
Penczek, P. see Bogdal, D.: Vol. 163, pp. 193–263.
Penelle, J., Hall, H. K., Padias, A. B. and *Tanaka, H.*: Captodative Olefins in Polymer Chemistry. Vol. 102, pp. 73–104.
Peppas, N. A. see Bell, C. L.: Vol. 122, pp. 125–176.
Peppas, N. A. see Hassan, C. M.: Vol. 153, pp. 37–65.
Peppas, N. A. see Narasimhan, B.: Vol. 128, pp. 157–208.
Petersen, K. L. see Geil, P. H.: Vol. 180, pp. 89–159.
Pet'ko, I. P. see Batog, A. E.: Vol. 144, pp. 49–114.
Pheyghambarian, N. see Kippelen, B.: Vol. 161, pp. 87–156.
Pichot, C. see Hunkeler, D.: Vol. 112, pp. 115–134.
Picken, S. J. see Northolt, M. G.: Vol. 178, pp. 1–108.
Pielichowski, J. see Bogdal, D.: Vol. 163, pp. 193–263.
Pielichowski, J. see Penczek, P.: Vol. 184, pp. 1–95.
Pieper, T. see Kilian, H. G.: Vol. 108, pp. 49–90.
Pispas, S. see Pitsikalis, M.: Vol. 135, pp. 1–138.
Pispas, S. see Hadjichristidis, N.: Vol. 142, pp. 71–128.
Pitsikalis, M., Pispas, S., Mays, J. W. and *Hadjichristidis, N.*: Nonlinear Block Copolymer Architectures. Vol. 135, pp. 1–138.
Pitsikalis, M. see Hadjichristidis, N.: Vol. 142, pp. 71–128.
Pitsikalis, M. see Hadjichristidis, N.: Vol. 189, pp. 1–124.
Pleul, D. see Spange, S.: Vol. 165, pp. 43–78.
Plummer, C. J. G.: Microdeformation and Fracture in Bulk Polyolefins. Vol. 169, pp. 75–119.
Pötschke, D. see Dingenouts, N.: Vol. 144, pp. 1–48.
Pokrovskii, V. N.: The Mesoscopic Theory of the Slow Relaxation of Linear Macromolecules. Vol. 154, pp. 143–219.
Pospíšil, J.: Functionalized Oligomers and Polymers as Stabilizers for Conventional Polymers. Vol. 101, pp. 65–168.
Pospíšil, J.: Aromatic and Heterocyclic Amines in Polymer Stabilization. Vol. 124, pp. 87–190.
Powers, A. C. see Prokop, A.: Vol. 136, pp. 53–74.
Prasad, P. N. see Lin, T.-C.: Vol. 161, pp. 157–193.
Priddy, D. B.: Recent Advances in Styrene Polymerization. Vol. 111, pp. 67–114.
Priddy, D. B.: Thermal Discoloration Chemistry of Styrene-co-Acrylonitrile. Vol. 121, pp. 123–154.
Privalko, V. P. and *Novikov, V. V.*: Model Treatments of the Heat Conductivity of Heterogeneous Polymers. Vol. 119, pp. 31–78.
Prociak, A. see Bogdal, D.: Vol. 163, pp. 193–263.
Prokop, A., Hunkeler, D., DiMari, S., Haralson, M. A. and *Wang, T. G.*: Water Soluble Polymers for Immunoisolation I: Complex Coacervation and Cytotoxicity. Vol. 136, pp. 1–52.
Prokop, A., Hunkeler, D., Powers, A. C., Whitesell, R. R. and *Wang, T. G.*: Water Soluble Polymers for Immunoisolation II: Evaluation of Multicomponent Microencapsulation Systems. Vol. 136, pp. 53–74.
Prokop, A., Kozlov, E., Carlesso, G. and *Davidsen, J. M.*: Hydrogel-Based Colloidal Polymeric System for Protein and Drug Delivery: Physical and Chemical Characterization, Permeability Control and Applications. Vol. 160, pp. 119–174.

Pruitt, L. A.: The Effects of Radiation on the Structural and Mechanical Properties of Medical Polymers. Vol. 162, pp. 65–95.
Pudavar, H. E. see Lin, T.-C.: Vol. 161, pp. 157–193.
Pukánszky, B. and *Fekete, E.*: Adhesion and Surface Modification. Vol. 139, pp. 109–154.
Putnam, D. and *Kopecek, J.*: Polymer Conjugates with Anticancer Acitivity. Vol. 122, pp. 55–124.
Putra, E. G. R. see Ungar, G.: Vol. 180, pp. 45–87.

Quirk, R. P., Yoo, T., Lee, Y., M., Kim, J. and *Lee, B.*: Applications of 1,1-Diphenylethylene Chemistry in Anionic Synthesis of Polymers with Controlled Structures. Vol. 153, pp. 67–162.

Ramaraj, R. and *Kaneko, M.*: Metal Complex in Polymer Membrane as a Model for Photosynthetic Oxygen Evolving Center. Vol. 123, pp. 215–242.
Rangarajan, B. see Scranton, A. B.: Vol. 122, pp. 1–54.
Ranucci, E. see Söderqvist Lindblad, M.: Vol. 157, pp. 139–161.
Raphaël, E. see Léger, L.: Vol. 138, pp. 185–226.
Rastogi, S. and *Terry, A. E.*: Morphological implications of the interphase bridging crystalline and amorphous regions in semi-crystalline polymers. Vol. 180, pp. 161–194.
Ray, C. R. and *Moore, J. S.*: Supramolecular Organization of Foldable Phenylene Ethynylene Oligomers. Vol. 177, pp. 99–149.
Reddinger, J. L. and *Reynolds, J. R.*: Molecular Engineering of p-Conjugated Polymers. Vol. 145, pp. 57–122.
Reghunadhan Nair, C. P., Mathew, D. and *Ninan, K. N.*: Cyanate Ester Resins, Recent Developments. Vol. 155, pp. 1–99.
Reichert, K. H. see Hunkeler, D.: Vol. 112, pp. 115–134.
Rehahn, M., Mattice, W. L. and *Suter, U. W.*: Rotational Isomeric State Models in Macromolecular Systems. Vol. 131/132, pp. 1–475.
Rehahn, M. see Bohrisch, J.: Vol. 165, pp. 1–41.
Rehahn, M. see Holm, C.: Vol. 166, pp. 1–27.
Reineker, P. see Holm, C.: Vol. 166, pp. 67–111.
Reitberger, T. see Jacobson, K.: Vol. 169, pp. 151–176.
Reynolds, J. R. see Reddinger, J. L.: Vol. 145, pp. 57–122.
Richter, D. see Ewen, B.: Vol. 134, pp. 1–130.
Richter, D., Monkenbusch, M. and *Colmenero, J.*: Neutron Spin Echo in Polymer Systems. Vol. 174, pp. 1–221.
Riegler, S. see Trimmel, G.: Vol. 176, pp. 43–87.
Ringsdorf, H. see Duncan, R.: Vol. 192, pp. 1–8.
Risse, W. see Grubbs, R.: Vol. 102, pp. 47–72.
Rivas, B. L. and *Geckeler, K. E.*: Synthesis and Metal Complexation of Poly(ethyleneimine) and Derivatives. Vol. 102, pp. 171–188.
Roberts, G. W. see Kennedy, K. A.: Vol. 175, pp. 329–346.
Robin, J. J.: The Use of Ozone in the Synthesis of New Polymers and the Modification of Polymers. Vol. 167, pp. 35–79.
Robin, J. J. see Boutevin, B.: Vol. 102, pp. 105–132.
Rodríguez-Pérez, M. A.: Crosslinked Polyolefin Foams: Production, Structure, Properties, and Applications. Vol. 184, pp. 97–126.
Roe, R.-J.: MD Simulation Study of Glass Transition and Short Time Dynamics in Polymer Liquids. Vol. 116, pp. 111–114.

Roovers, J. and *Comanita, B.*: Dendrimers and Dendrimer-Polymer Hybrids. Vol. 142, pp. 179–228.
Rothon, R. N.: Mineral Fillers in Thermoplastics: Filler Manufacture and Characterisation. Vol. 139, pp. 67–108.
de Rosa, C. see Auriemma, F.: Vol. 181, pp. 1–74.
Rozenberg, B. A. see Williams, R. J. J.: Vol. 128, pp. 95–156.
Rühe, J., Ballauff, M., Biesalski, M., Dziezok, P., Gröhn, F., Johannsmann, D., Houbenov, N., Hugenberg, N., Konradi, R., Minko, S., Motornov, M., Netz, R. R., Schmidt, M., Seidel, C., Stamm, M., Stephan, T., Usov, D. and *Zhang, H.*: Polyelectrolyte Brushes. Vol. 165, pp. 79–150.
Ruckenstein, E.: Concentrated Emulsion Polymerization. Vol. 127, pp. 1–58.
Ruiz-Taylor, L. see Mathieu, H. J.: Vol. 162, pp. 1–35.
Rusanov, A. L.: Novel Bis (Naphtalic Anhydrides) and Their Polyheteroarylenes with Improved Processability. Vol. 111, pp. 115–176.
Rusanov, A. L., Likhatchev, D., Kostoglodov, P. V., Müllen, K. and *Klapper, M.*: Proton-Exchanging Electrolyte Membranes Based on Aromatic Condensation Polymers. Vol. 179, pp. 83–134.
Russel, T. P. see Hedrick, J. L.: Vol. 141, pp. 1–44.
Russum, J. P. see Schork, F. J.: Vol. 175, pp. 129–255.
Rychly, J. see Lazár, M.: Vol. 102, pp. 189–222.
Ryner, M. see Stridsberg, K. M.: Vol. 157, pp. 27–51.
Ryzhov, V. A. see Bershtein, V. A.: Vol. 114, pp. 43–122.

Sabsai, O. Y. see Barshtein, G. R.: Vol. 101, pp. 1–28.
Saburov, V. V. see Zubov, V. P.: Vol. 104, pp. 135–176.
Saito, S., Konno, M. and *Inomata, H.*: Volume Phase Transition of N-Alkylacrylamide Gels. Vol. 109, pp. 207–232.
Samsonov, G. V. and *Kuznetsova, N. P.*: Crosslinked Polyelectrolytes in Biology. Vol. 104, pp. 1–50.
Santa Cruz, C. see Baltá-Calleja, F. J.: Vol. 108, pp. 1–48.
Santos, S. see Baschnagel, J.: Vol. 152, p. 41–156.
Satchi-Fainaro, R. see Duncan, R.: Vol. 192, pp. 1–8.
Sato, T. and *Teramoto, A.*: Concentrated Solutions of Liquid-Christalline Polymers. Vol. 126, pp. 85–162.
Schaller, C. see Bohrisch, J.: Vol. 165, pp. 1–41.
Schäfer, R. see Köhler, W.: Vol. 151, pp. 1–59.
Scherf, U. and *Müllen, K.*: The Synthesis of Ladder Polymers. Vol. 123, pp. 1–40.
Schlatmann, R. see Northolt, M. G.: Vol. 178, pp. 1–108.
Schmid, F. see Müller, M.: Vol. 185, pp. 1–58.
Schmidt, M. see Förster, S.: Vol. 120, pp. 51–134.
Schmidt, M. see Rühe, J.: Vol. 165, pp. 79–150.
Schmidt, M. see Volk, N.: Vol. 166, pp. 29–65.
Scholz, M.: Effects of Ion Radiation on Cells and Tissues. Vol. 162, pp. 97–158.
Schönherr, H. see Vancso, G. J.: Vol. 182, pp. 55–129.
Schopf, G. and *Koßmehl, G.*: Polythiophenes – Electrically Conductive Polymers. Vol. 129, pp. 1–145.
Schork, F. J., Luo, Y., Smulders, W., Russum, J. P., Butté, A. and *Fontenot, K.*: Miniemulsion Polymerization. Vol. 175, pp. 127–255.
Schulz, E. see Munz, M.: Vol. 164, pp. 97–210.
Schwahn, D.: Critical to Mean Field Crossover in Polymer Blends. Vol. 183, pp. 1–61.

Seppälä, J. see *Löfgren, B.*: Vol. 169, pp. 1–12.
Sturm, H. see *Munz, M.*: Vol. 164, pp. 87–210.
Schweizer, K. S.: Prism Theory of the Structure, Thermodynamics, and Phase Transitions of Polymer Liquids and Alloys. Vol. 116, pp. 319–378.
Scranton, A. B., Rangarajan, B. and *Klier, J.*: Biomedical Applications of Polyelectrolytes. Vol. 122, pp. 1–54.
Sefton, M. V. and *Stevenson, W. T. K.*: Microencapsulation of Live Animal Cells Using Polycrylates. Vol. 107, pp. 143–198.
Seidel, C. see *Holm, C.*: Vol. 166, pp. 67–111.
Seidel, C. see *Rühe, J.*: Vol. 165, pp. 79–150.
El Seoud, O. A. and *Heinze, T.*: Organic Esters of Cellulose: New Perspectives for Old Polymers. Vol. 186, pp. 103–149.
Shabat, D. see *Amir, R. J.*: Vol. 192, pp. 59–94.
Shamanin, V. V.: Bases of the Axiomatic Theory of Addition Polymerization. Vol. 112, pp. 135–180.
Shcherbina, M. A. see *Ungar, G.*: Vol. 180, pp. 45–87.
Sheiko, S. S.: Imaging of Polymers Using Scanning Force Microscopy: From Superstructures to Individual Molecules. Vol. 151, pp. 61–174.
Sherrington, D. C. see *Cameron, N. R.*: Vol. 126, pp. 163–214.
Sherrington, D. C. see *Lin, J.*: Vol. 111, pp. 177–220.
Sherrington, D. C. see *Steinke, J.*: Vol. 123, pp. 81–126.
Shibayama, M. see *Tanaka, T.*: Vol. 109, pp. 1–62.
Shiga, T.: Deformation and Viscoelastic Behavior of Polymer Gels in Electric Fields. Vol. 134, pp. 131–164.
Shim, H.-K. and *Jin, J.*: Light-Emitting Characteristics of Conjugated Polymers. Vol. 158, pp. 191–241.
Shoda, S. see *Kobayashi, S.*: Vol. 121, pp. 1–30.
Siegel, R. A.: Hydrophobic Weak Polyelectrolyte Gels: Studies of Swelling Equilibria and Kinetics. Vol. 109, pp. 233–268.
de Silva, D. S. M. see *Ungar, G.*: Vol. 180, pp. 45–87.
Silvestre, F. see *Calmon-Decriaud, A.*: Vol. 207, pp. 207–226.
Sillion, B. see *Mison, P.*: Vol. 140, pp. 137–180.
Simon, F. see *Spange, S.*: Vol. 165, pp. 43–78.
Simon, G. P. see *Becker, O.*: Vol. 179, pp. 29–82.
Simon, P. F. W. see *Abetz, V.*: Vol. 189, pp. 125–212.
Simonutti, R. see *Sozzani, P.*: Vol. 181, pp. 153–177.
Singh, R. P. see *Sivaram, S.*: Vol. 101, pp. 169–216.
Singh, R. P. see *Desai, S. M.*: Vol. 169, pp. 231–293.
Sinha Ray, S. see *Biswas, M.*: Vol. 155, pp. 167–221.
Sivaram, S. and *Singh, R. P.*: Degradation and Stabilization of Ethylene-Propylene Copolymers and Their Blends: A Critical Review. Vol. 101, pp. 169–216.
Slugovc, C. see *Trimmel, G.*: Vol. 176, pp. 43–87.
Smulders, W. see *Schork, F. J.*: Vol. 175, pp. 129–255.
Soares, J. B. P. see *Anantawaraskul, S.*: Vol. 182, pp. 1–54.
Sozzani, P., Bracco, S., Comotti, A. and *Simonutti, R.*: Motional Phase Disorder of Polymer Chains as Crystallized to Hexagonal Lattices. Vol. 181, pp. 153–177.
Söderqvist Lindblad, M., Liu, Y., Albertsson, A.-C., Ranucci, E. and *Karlsson, S.*: Polymer from Renewable Resources. Vol. 157, pp. 139–161.
Spange, S., Meyer, T., Voigt, I., Eschner, M., Estel, K., Pleul, D. and *Simon, F.*: Poly(Vinylformamide-co-Vinylamine)/Inorganic Oxid Hybrid Materials. Vol. 165, pp. 43–78.

Stamm, M. see Möhwald, H.: Vol. 165, pp. 151–175.
Stamm, M. see Rühe, J.: Vol. 165, pp. 79–150.
Starodybtzev, S. see Khokhlov, A.: Vol. 109, pp. 121–172.
Stegeman, G. I. see Canva, M.: Vol. 158, pp. 87–121.
Steinke, J., Sherrington, D. C. and *Dunkin, I. R.*: Imprinting of Synthetic Polymers Using Molecular Templates. Vol. 123, pp. 81–126.
Stelzer, F. see Trimmel, G.: Vol. 176, pp. 43–87.
Stenberg, B. see Jacobson, K.: Vol. 169, pp. 151–176.
Stenzenberger, H. D.: Addition Polyimides. Vol. 117, pp. 165–220.
Stephan, T. see Rühe, J.: Vol. 165, pp. 79–150.
Stevenson, W. T. K. see Sefton, M. V.: Vol. 107, pp. 143–198.
Stridsberg, K. M., Ryner, M. and *Albertsson, A.-C.*: Controlled Ring-Opening Polymerization: Polymers with Designed Macromoleculars Architecture. Vol. 157, pp. 27–51.
Sturm, H. see Munz, M.: Vol. 164, pp. 87–210.
Suematsu, K.: Recent Progress of Gel Theory: Ring, Excluded Volume, and Dimension. Vol. 156, pp. 136–214.
Sugimoto, H. and *Inoue, S.*: Polymerization by Metalloporphyrin and Related Complexes. Vol. 146, pp. 39–120.
Suginome, M. and *Ito, Y.*: Transition Metal-Mediated Polymerization of Isocyanides. Vol. 171, pp. 77–136.
Sumpter, B. G., Noid, D. W., Liang, G. L. and *Wunderlich, B.*: Atomistic Dynamics of Macromolecular Crystals. Vol. 116, pp. 27–72.
Sumpter, B. G. see Otaigbe, J. U.: Vol. 154, pp. 1–86.
Sun, H.-B. and *Kawata, S.*: Two-Photon Photopolymerization and 3D Lithographic Microfabrication. Vol. 170, pp. 169–273.
Suter, U. W. see Gusev, A. A.: Vol. 116, pp. 207–248.
Suter, U. W. see Leontidis, E.: Vol. 116, pp. 283–318.
Suter, U. W. see Rehahn, M.: Vol. 131/132, pp. 1–475.
Suter, U. W. see Baschnagel, J.: Vol. 152, pp. 41–156.
Suzuki, A.: Phase Transition in Gels of Sub-Millimeter Size Induced by Interaction with Stimuli. Vol. 110, pp. 199–240.
Suzuki, A. and *Hirasa, O.*: An Approach to Artifical Muscle by Polymer Gels due to Micro-Phase Separation. Vol. 110, pp. 241–262.
Suzuki, K. see Nomura, M.: Vol. 175, pp. 1–128.
Swiatkiewicz, J. see Lin, T.-C.: Vol. 161, pp. 157–193.

Tagawa, S.: Radiation Effects on Ion Beams on Polymers. Vol. 105, pp. 99–116.
Taguet, A., Ameduri, B. and *Boutevin, B.*: Crosslinking of Vinylidene Fluoride-Containing Fluoropolymers. Vol. 184, pp. 127–211.
Takata, T., Kihara, N. and *Furusho, Y.*: Polyrotaxanes and Polycatenanes: Recent Advances in Syntheses and Applications of Polymers Comprising of Interlocked Structures. Vol. 171, pp. 1–75.
Takeuchi, D. see Osakada, K.: Vol. 171, pp. 137–194.
Tan, K. L. see Kang, E. T.: Vol. 106, pp. 135–190.
Tanaka, H. and *Shibayama, M.*: Phase Transition and Related Phenomena of Polymer Gels. Vol. 109, pp. 1–62.
Tanaka, T. see Penelle, J.: Vol. 102, pp. 73–104.
Tauer, K. see Guyot, A.: Vol. 111, pp. 43–66.
Teramoto, A. see Sato, T.: Vol. 126, pp. 85–162.

Terent'eva, J. P. and *Fridman, M. L.*: Compositions Based on Aminoresins. Vol. 101, pp. 29–64.
Terry, A. E. see Rastogi, S.: Vol. 180, pp. 161–194.
Theodorou, D. N. see Dodd, L. R.: Vol. 116, pp. 249–282.
Thomson, R. C., Wake, M. C., Yaszemski, M. J. and *Mikos, A. G.*: Biodegradable Polymer Scaffolds to Regenerate Organs. Vol. 122, pp. 245–274.
Thünemann, A. F., Müller, M., Dautzenberg, H., Joanny, J.-F. and *Löwen, H.*: Polyelectrolyte complexes. Vol. 166, pp. 113–171.
Tieke, B. see v. Klitzing, R.: Vol. 165, pp. 177–210.
Tobita, H. see Nomura, M.: Vol. 175, pp. 1–128.
Tokita, M.: Friction Between Polymer Networks of Gels and Solvent. Vol. 110, pp. 27–48.
Traser, S. see Bohrisch, J.: Vol. 165, pp. 1–41.
Tries, V. see Baschnagel, J.: Vol. 152, p. 41–156.
Trimmel, G., Riegler, S., Fuchs, G., Slugovc, C. and *Stelzer, F.*: Liquid Crystalline Polymers by Metathesis Polymerization. Vol. 176, pp. 43–87.
Tsuruta, T.: Contemporary Topics in Polymeric Materials for Biomedical Applications. Vol. 126, pp. 1–52.

Uemura, T., Naka, K. and *Chujo, Y.*: Functional Macromolecules with Electron-Donating Dithiafulvene Unit. Vol. 167, pp. 81–106.
Ungar, G., Putra, E. G. R., de Silva, D. S. M., Shcherbina, M. A. and *Waddon, A. J.*: The Effect of Self-Poisoning on Crystal Morphology and Growth Rates. Vol. 180, pp. 45–87.
Usov, D. see Rühe, J.: Vol. 165, pp. 79–150.
Usuki, A., Hasegawa, N. and *Kato, M.*: Polymer-Clay Nanocomposites. Vol. 179, pp. 135–195.
Uyama, H. see Kobayashi, S.: Vol. 121, pp. 1–30.
Uyama, Y.: Surface Modification of Polymers by Grafting. Vol. 137, pp. 1–40.

Vancso, G. J., Hillborg, H. and *Schönherr, H.*: Chemical Composition of Polymer Surfaces Imaged by Atomic Force Microscopy and Complementary Approaches. Vol. 182, pp. 55–129.
Varma, I. K. see Albertsson, A.-C.: Vol. 157, pp. 99–138.
Vasilevskaya, V. see Khokhlov, A.: Vol. 109, pp. 121–172.
Vaskova, V. see Hunkeler, D.: Vol. 112, pp. 115–134.
Verdugo, P.: Polymer Gel Phase Transition in Condensation-Decondensation of Secretory Products. Vol. 110, pp. 145–156.
Veronese, F. M. see Pasut, G.: Vol. 192, pp. 95–134.
Vettegren, V. I. see Bronnikov, S. V.: Vol. 125, pp. 103–146.
Vilgis, T. A. see Holm, C.: Vol. 166, pp. 67–111.
Viovy, J.-L. and *Lesec, J.*: Separation of Macromolecules in Gels: Permeation Chromatography and Electrophoresis. Vol. 114, pp. 1–42.
Vlahos, C. see Hadjichristidis, N.: Vol. 142, pp. 71–128.
Voigt, I. see Spange, S.: Vol. 165, pp. 43–78.
Volk, N., Vollmer, D., Schmidt, M., Oppermann, W. and *Huber, K.*: Conformation and Phase Diagrams of Flexible Polyelectrolytes. Vol. 166, pp. 29–65.
Volksen, W.: Condensation Polyimides: Synthesis, Solution Behavior, and Imidization Characteristics. Vol. 117, pp. 111–164.
Volksen, W. see Hedrick, J. L.: Vol. 141, pp. 1–44.
Volksen, W. see Hedrick, J. L.: Vol. 147, pp. 61–112.
Vollmer, D. see Volk, N.: Vol. 166, pp. 29–65.
Voskerician, G. and *Weder, C.*: Electronic Properties of PAEs. Vol. 177, pp. 209–248.

Waddon, A. J. see Ungar, G.: Vol. 180, pp. 45–87.
Wagener, K. B. see Baughman, T. W.: Vol. 176, pp. 1–42.
Wagner, E. and *Kloeckner, J.*: Gene Delivery Using Polymer Therapeutics. Vol. 192, pp. 135–173.
Wake, M. C. see Thomson, R. C.: Vol. 122, pp. 245–274.
Wandrey, C., Hernández-Barajas, J. and *Hunkeler, D.*: Diallyldimethylammonium Chloride and its Polymers. Vol. 145, pp. 123–182.
Wang, K. L. see Cussler, E. L.: Vol. 110, pp. 67–80.
Wang, S.-Q.: Molecular Transitions and Dynamics at Polymer/Wall Interfaces: Origins of Flow Instabilities and Wall Slip. Vol. 138, pp. 227–276.
Wang, S.-Q. see Bhargava, R.: Vol. 163, pp. 137–191.
Wang, T. G. see Prokop, A.: Vol. 136, pp. 1–52; 53–74.
Wang, X. see Lin, T.-C.: Vol. 161, pp. 157–193.
Watanabe, K. see Hikosaka, M.: Vol. 191, pp. 137–186.
Webster, O. W.: Group Transfer Polymerization: Mechanism and Comparison with Other Methods of Controlled Polymerization of Acrylic Monomers. Vol. 167, pp. 1–34.
Weder, C. see Voskerician, G.: Vol. 177, pp. 209–248.
Weis, J.-J. and *Levesque, D.*: Simple Dipolar Fluids as Generic Models for Soft Matter. Vol. 185, pp. 163–225.
Whitesell, R. R. see Prokop, A.: Vol. 136, pp. 53–74.
Williams, R. A. see Geil, P. H.: Vol. 180, pp. 89–159.
Williams, R. J. J., Rozenberg, B. A. and *Pascault, J.-P.*: Reaction Induced Phase Separation in Modified Thermosetting Polymers. Vol. 128, pp. 95–156.
Winkler, R. G. see Holm, C.: Vol. 166, pp. 67–111.
Winter, H. H. and *Mours, M.*: Rheology of Polymers Near Liquid-Solid Transitions. Vol. 134, pp. 165–234.
Wittmeyer, P. see Bohrisch, J.: Vol. 165, pp. 1–41.
Wood-Adams, P. M. see Anantawaraskul, S.: Vol. 182, pp. 1–54.
Wu, C.: Laser Light Scattering Characterization of Special Intractable Macromolecules in Solution. Vol. 137, pp. 103–134.
Wunderlich, B. see Sumpter, B. G.: Vol. 116, pp. 27–72.

Xiang, M. see Jiang, M.: Vol. 146, pp. 121–194.
Xie, T. Y. see Hunkeler, D.: Vol. 112, pp. 115–134.
Xu, P. see Geil, P. H.: Vol. 180, pp. 89–159.
Xu, Z., Hadjichristidis, N., Fetters, L. J. and *Mays, J. W.*: Structure/Chain-Flexibility Relationships of Polymers. Vol. 120, pp. 1–50.

Yagci, Y. and *Endo, T.*: N-Benzyl and N-Alkoxy Pyridium Salts as Thermal and Photochemical Initiators for Cationic Polymerization. Vol. 127, pp. 59–86.
Yamaguchi, I. see Yamamoto, T.: Vol. 177, pp. 181–208.
Yamamoto, T.: Molecular Dynamics Modeling of the Crystal-Melt Interfaces and the Growth of Chain Folded Lamellae. Vol. 191, pp. 37–85.
Yamamoto, T., Yamaguchi, I. and *Yasuda, T.*: PAEs with Heteroaromatic Rings. Vol. 177, pp. 181–208.
Yamaoka, H.: Polymer Materials for Fusion Reactors. Vol. 105, pp. 117–144.
Yamazaki, S. see Hikosaka, M.: Vol. 191, pp. 137–186.
Yannas, I. V.: Tissue Regeneration Templates Based on Collagen-Glycosaminoglycan Copolymers. Vol. 122, pp. 219–244.
Yang, J. see Geil, P. H.: Vol. 180, pp. 89–159.

Yang, J. S. see Jo, W. H.: Vol. 156, pp. 1–52.
Yasuda, H. and *Ihara, E.*: Rare Earth Metal-Initiated Living Polymerizations of Polar and Nonpolar Monomers. Vol. 133, pp. 53–102.
Yasuda, T. see Yamamoto, T.: Vol. 177, pp. 181–208.
Yaszemski, M. J. see Thomson, R. C.: Vol. 122, pp. 245–274.
Yoo, T. see Quirk, R. P.: Vol. 153, pp. 67–162.
Yoon, D. Y. see Hedrick, J. L.: Vol. 141, pp. 1–44.
Yoshida, H. and *Ichikawa, T.*: Electron Spin Studies of Free Radicals in Irradiated Polymers. Vol. 105, pp. 3–36.

Zhang, H. see Rühe, J.: Vol. 165, pp. 79–150.
Zhang, Y.: Synchrotron Radiation Direct Photo Etching of Polymers. Vol. 168, pp. 291–340.
Zheng, J. and *Swager, T. M.*: Poly(arylene ethynylene)s in Chemosensing and Biosensing. Vol. 177, pp. 151–177.
Zhou, H. see Jiang, M.: Vol. 146, pp. 121–194.
Zhou, Z. see Abe, A.: Vol. 181, pp. 121–152.
Zubov, V. P., Ivanov, A. E. and *Saburov, V. V.*: Polymer-Coated Adsorbents for the Separation of Biopolymers and Particles. Vol. 104, pp. 135–176.

Subject Index

Active targeting *II* 7–9
Adamantane *I* 145
Adaptor units, De Groot *I* 69
– McGrath *I* 70
– Shabat *I* 68
Alcohol dehydrogenase *I* 99
2-(4-Aminobenzylidene propane)-1,3-diol *I* 69
Aminomethyl-pyrene *I* 72
Aminopropyltriethoxysilane *II* 135
Amyloid formation *II* 160–162
Angiogenesis *II* 9–11
– drug targeting *II* 48–49
– markers *II* 43–45
Angiogenesis inhibitors *II* 9–11, 22
Angiotensin-induced hypertension *II* 108–109
Anthrax toxin, sequestration *I* 36
Anti-anthrax agents *I* 36
Antibodies, PEGylation *I* 118, 124
Antibody-G5 PAMAM *I* 63
Anti-HIV agents, anionic polymers *I* 40
Anti-infective polymeric drugs *I* 30
Antimicrobial agents, polyvalent *I* 40
Anti-obesity drugs *I* 47
Antiviral agents, polyvalent ligands *I* 37
Arginine deaminase, PEGylation *I* 121
Arginine-grafted dendrimer *I* 149
Artificial vaccines *II* 173–198
Asialoglycoprotein *I* 140
Atomic force microscopy *II* 123–172
– amyloid formation *II* 160–162
– contact mode *II* 128–129
– cryogenic *II* 130
– DNA *II* 134–146
– force-displacement measurements *II* 132–134
– imaging proteins on substrates *II* 157–160

– instrumentation *II* 126–128
– liquid imaging *II* 130–131
– membrane proteins *II* 154–155
– nontopographical applications *II* 131–132
– phase imaging *II* 129–130
– polysaccharides *II* 150–153
– RNA *II* 146–150
– single molecule biopolymers *II* 134–146
– tapping mode *II* 129
– tip geometry and carbon nanotubes *II* 131
– viral proteins *II* 156–157
ATP binding cassette proteins *II* 175

Bacillus anthracis *I* 36
Bacterial toxins *I* 32
Bile acid sequestrants *I* 25, 28
Bioavailability, oral *I* 12
Bioconjugates, PEG *I* 95
Biopolymers *II* 134–146
– DNA condensation *II* 137–138
– DNA immobilisation *II* 134–137
– DNA-drug interactions *II* 142–144
– DNA-protein interactions *II* 138–142
Bioresponsive polymers *I* 164
2,4-Bis(hydroxymethyl)phenol *I* 70
2,6-Bishydroxymethyl-*p*-cresol *I* 68
Block copolymers *II* 69–70
– synthesis *II* 72–76
Bradykinin *II* 109–110
Brain tumour implants *II* 38–39
Breast cancer resistance protein *II* 175
Bystander effect *II* 25

Camptothecin *I* 79
Camptothecin conjugates *II* 35–36, 116
Cancer chemotherapy *II* 103–121

Carboxylic groups, PEGylation *I* 127
Carboxymethyl dextran-CPT II analogues 39–40
Catalase *II* 116
Catecholates *I* 23
CDDP-incorporated micelles *II* 3–4
Cellulose *II* 150
Chitosan *I* 143
Cholesterol, LDLc *I* 25
Cholesterol-lowering drugs *I* 25
Cholestyramine *I* 26, 33
Clostridium difficile toxin, sequestration *I* 32
Colestipol *I* 26, 33
Coronary heart disease *I* 25
Critical micelle concentration *II* 76
Cyclodextrin dendrimer *I* 66
Cyclodextrins *I* 144, 148

DAB-64 *I* 64
DE-310 *II* 39
De Groot adaptor unit *I* 69
Dendrimers *I* 59
– cascade-release *I* 72
– self-immolative *I* 72
Dendrons, domino *I* 67
– multi-triggered *I* 89
Desferrioxamine *I* 22
Dextran *II* 152–153
Dextran-doxorubicin *II* 37
Diaminobutane poly(propylene imine) *I* 64
Diarrhea *I* 32
Divalent cations *II* 135–136
Divinylether-maleic anhydride *I* 3
DNA *II* 134–138
DNA condensation *II* 137–138
DNA crosslinkers *II* 142–143
DNA delivery *I* 135
DNA-drug interactions *II* 142–144
– direct imaging *II* 145–146
– DNA crosslinkers *II* 142–143
– intercalators *II* 143–144
– minor groove binders *II* 144
DNA immobilisation *II* 134–137
– divalent cations *II* 135–136
– monovalent ions *II* 136–137
– silanes *II* 135
DNA microarrays *II* 180
DNA polyplexes, caged *I* 156

DNA-protein interactions *II* 138–142
Domino dendrimers *I* 59, 71
Domino dendrons, multi-triggered *I* 88
Doxorubicin *I* 4, 79
Drug delivery systems *II* 1–65, 71–72
– polyion complex micelles *II* 88–95
– polymer vesicles *II* 95–96
– polymeric micelles *see* Polymeric micelles
Drug resistance *II* 173–198
Drug targeting *II* 1–65, 67–101
– cancer chemotherapy *II* 103–121
– parenteral *II* 11
– passive versus active *II* 7–9
– tumour cells versus tumour vasculature *II* 9–11
– tumour vasculature *II* 42–50

Electrolyte homeostasis *I* 14
Endotoxins *I* 32
Enhanced permeability and retention effect *II* 7, 8, 23, 72, 103–121
– modulation of *II* 108–110
– theory and principles *II* 106–108
EPR effect *I* 4
Erythropoietin, PEGylation *I* 121
Exotoxins *I* 32

Fat binder, polymeric *I* 49
Fenton reaction *I* 21

Gene delivery *I* 135, *II* 90–95
– device *I* 62
Gene expression profiles *II* 183–185
– transcriptional activation by synthetic polymers *II* 185–189
Gene therapy *I* 5
Genomic profiles *II* 180–183
Glucosamine dendrimer *I* 66
Glycocholate 26
Granulocyte colony stimulating factor *I* 112
Growth hormone *I* 115

Hemagglutination *I* 37
Hemochromatosis *I* 22
Hemoglobin *I* 52
Heparanproteoglycans *I* 141
HIV virus *I* 40
HK polymer *I* 139

Subject Index

HMG-CoA reductase inhibitors *I* 25
HPMA *I* 4
HPMA copolymer *II* 1–65
HPMA copolymer-1,5-diazaanthraquinone *II* 42
HPMA copolymer-antibody-doxorubicin conjugates *II* 26–27
HPMA copolymer-camptothecin *II* 28–31
HPMA copolymer-DACH platinate *II* 32–33
HPMA copolymer-doxorubicin-galactosamine *II* 22
HPMA copolymer-Gly-Phe-Leu-Gly-doxorubicin *II* 23–25
HPMA copolymer-Gly-Phe-Leu-Gly-doxorubicin-galactosamine *II* 25–26
HPMA copolymer-paclitaxel *II* 27–28
HPMA copolymer-platinate *II* 31–32
HPMA copolymer-TNP-470 (caplostatin) *II* 46–48
Human groth hormone *I* 115
Hydrazones *I* 154
Hydrophobic drugs, encapsulation of *II* 80–83
Hydroxamates *I* 23
Hydroxypropyl methacrylate *I* 159
Hyperkalemia *I* 14
Hyperphosphatemia *I* 15

IgG *I* 119
Immune response *II* 189–191
Immunocamouflage *I* 129
Inflammatory mediators *II* 110
Influenza virus inhibitors, polyvalent *I* 38
Inorganic ions *I* 14
Insulin, PEGylation *I* 120
Intercalators *II* 143–144
– direct imaging *II* 145–146
Interferons *I* 105, 109
Iron, sequestration *I* 21
Iron overload disorder *I* 21

LDLc *I* 25
Lectin *I* 67
Leptin (OB proteins), PEGylation *I* 121
Leukemia *I* 84
Ligand-receptor interactions *II* 164–166

Ligands *I* 31
Lipase inhibition *I* 48
Lipoprotein cholesterol *I* 25
Liposomal-PEG-Ala-Pro-Arg-Pro-Gly *II* 49–50
Lysosomotropic drug delivery *II* 19

Macromolecular drugs *II* 103–121
– intracellular uptake *II* 111
– quality of life *II* 114–115
– SMANCS *II* 111–113
McGrath adaptor unit *I* 70
Megakaryocyte growth and development factor *I* 114
Membrane proteins *II* 154–155
Mescaline-*N*-vinylpyrolidone *I* 3
Metronidazole *I* 32
Micelle *I* 2
Minor groove binders *II* 144
MOLT-3 leukemia *I* 84
Monoclonal antibodies *II* 18
Monovalent ions *II* 136–137
MRSA *I* 41
Multidrug resistance *I* 40
Multidrug resistance-associated proteins *II* 175
Multiple sclerosis *I* 44
Multi-prodrug *I* 59
Myelin based protein *I* 45

Naphthalenesulfonate-formaldehyde, HIV *I* 40
N-(2-hydroxypropyl)methacrylamide *see* HPMA
New chemical entities *II* 6
4-Nitroaniline *I* 72
Nucleic acids, delivery *I* 135

Obesity *I* 47
Oligolysines, disulfide cross-linking *I* 156
Orlistat *I* 47, 48
Oxidation therapy *II* 117

PA63 *I* 36
Paclitaxel conjugates *II* 33–35, 115
PAMAM *I* 60
Parenteral drug targeting *II* 11
Passive targeting *II* 7–9
PEGylated proteins *I* 4, 95
Phosphate binder therapy *I* 15

Phosphate ions, sequestration *I* 15
PEI-cholesterol *I* 148
Penicillin-G-amidase *I* 88
PEO-PPO-PEO *I* 145
Peptides, PEGylated *I* 95
P-glycoprotein *II* 175
PEG *II* 1–65
PEG-adenosine deaminase *II* 13–14
PEG-camptothecin *II* 36
PEG-DAC *II* 55–56
PEG-DAO *II* 54–55, 118
PEG-granulocyte-colony stimulating factor *II* 15–16
PEG-interferon-α *II* 16–17
PEG-*L*-asparaginase *II* 7, 14–15
PEG-paclitaxel *II* 37
PEG-XO *II* 54, 118
PEG-ZnPP *II* 55–56, 118
PEGylated-liposome-Raf mutant *II* 49
PEGylated-liposomes *II* 49–50
– liposomal-PEG-Ala-Pro-Arg-Pro-Gly *II* 49–50
– PEGylated-liposome-Raf mutant *II* 49
Peptide motifs *II* 45–46
Phenotype *II* 175–179
Phenotypic correction of immune response *II* 189–191
Plasma half-life *II* 103–121
Pluronic F68 *I* 52
Pluronic block copolymers *II* 176–177
Polyacetal-diethylstilboestrol *II* 40–42
Poly(amidoamine), dendrimers *I* 60, 142, 149, 151
Polyamines *I* 149
Poly(β-amino ester) *I* 152
Polycations, polyplexes *I* 140
Polydispersity *I* 12
Polyethyleneimines (PEI) *I* 62, 141, 147
– cyclodextrin-grafted *I* 148
– dextran-grafted *I* 147
– PEG-grafted *I* 147
Poly(ethyleneglycol) *I* 4, 62, 98, 145
Poly(ethylene oxide) *II* 71
Polyion complex micelles *II* 88–95
– gene delivery *II* 90–95
– properties of *II* 88–89
Poly(glutamic acid) *I* 4
Polyglycerol dendrimers *I* 63
Poly(lactide) *II* 73
Polylactide-hydroxyproline *I* 146

Poly-L-glutamic acid conjugates *II* 115–119
– camptothecin *II* 35–36, 116
– paclitaxel *II* 33–35, 115
Poly(lys-(AEDTP)) *I* 157
Polylysines *I* 65, 140, 146, 149
Polymer directed enzyme prodrug therapy *II* 50–52
Polymer-DNA complex *I* 2
Polymer-drug conjugate *I* 2
Polymer genomics *II* 192–194
Polymeric micelle *I* 2
Polymer-protein conjugate *I* 2
Polymer therapeutics *II* 1–65
– combinations *II* 50–56
– *see also individual compounds*
Polymer vesicles *II* 95–96
Polymer-coated surfaces *II* 191–192
Polymer-drug conjugates *II* 18–39
– alteration of signal transduction *II* 179–180
– angiogenesis inhibitors *II* 22
– anticancer agents *II* 20–21
– brain tumour implants *II* 38–39
– dextran-doxorubicin *II* 37
– and gene expression profiles *II* 183–185
– and genomic profiles *II* 180–183
– HPMA copolymer-antibody-doxorubicin conjugates *II* 26–27
– HPMA copolymer-camptothecin *II* 28–31
– HPMA copolymer-DACH platinate *II* 32–33
– HPMA copolymer-Gly-Phe-Leu-Gly-doxorubicin *II* 23–25
– HPMA copolymer-Gly-Phe-Leu-Gly-doxorubicin-galactosamine *II* 25–26
– HPMA copolymer-paclitaxel *II* 27–28
– HPMA copolymer-platinate *II* 31–32
– PEG-camptothecin *II* 36, 116
– PEG-paclitaxel *II* 37, 115
– phenotypic selectivity *II* 175–179
– poly-L-glutamic(PG)-camptothecin *II* 35–36, 116
– poly-L-glutamic(PG)-paclitaxel *II* 33–35, 116
– polymeric micelles *II* 37–38

Polymer-enzyme liposome therapy
  II 52–54
Polymer-protein conjugates *II* 11–18
– PEG-adenosine deaminase *II* 13–14
– PEG-granulocyte-colony stimulating
  factor *II* 15–16
– PEG-interferon-α *II* 16–17
– PEG-L-asparaginase *II* 14–15
– preclinical *II* 18
– styrene-co-maleic
  anhydride-neocarzinostatin *II* 17–18
Polymeric micelles *II* 37–38, 69–70
– blood circulation and tissue distribution
  II 78–80
– CDDP-incorporated micelles *II* 83–84
– drug delivery *II* 80–88
– encapsulation of hydrophobic drugs
  II 80–83
– intracellular location *II* 87–88
– properties of *II* 76–77
– stimuli-triggered drug release *II* 84–85
– surface-functionalized *II* 86–87
Polymethacrylates *I* 143
Polyorthoesters *I* 154
Polyphosphoesters *I* 154
Polyphosphoramidates *I* 154
Polyplexes *I* 2, 135, 140
– bioresponsive *I* 164
Polypropylene oxide, SCA *I* 52
Polypropyleneimines (PPI), dendritic
  I 150
Polysaccharide-enzyme interactions
  II 153
Polysaccharides *II* 150–153
– cellulose *II* 150
– dextran *II* 152–153
– starch grains *II* 150–152
Poly(styrene-4-sulfonate), HIV *I* 40
Poly(styrene sulfonic acid) *I* 34
Polyvalency *I* 12
Polyvalent interactions *I* 30
Potassium, sequestration *I* 14
Prodrug, dendritic *I* 78
Propranolol *I* 61
Prostacyclin antagonists *II* 110
Protein-DNA interactions *II* 138–142
Proteins
– DNA-protein interactions *II* 138–142
– mechanical properties *II* 162–164
– membrane *II* 154–155

– polymer-protein conjugates *II* 11–18
– single-force molecular interactions
  II 162
– on substrates *II* 157–160
– viral *II* 156–157
Proton sponge *I* 141, 146
Pyran copolymer *I* 3

Quality of life *II* 114–115
Quinone methide *I* 70

RBCs, PEGylation *I* 129
Reactive oxygen species *II* 54, 103–121
Receptors *I* 31
Renal failure *I* 15
Rheumatoid arthritis *I* 44
RNA *II* 146–150
– crystallization *II* 149
– force investigations *II* 149–150
– in situ synthesis *II* 147
– tectonics *II* 147–149
Rotavirus *I* 37

Sequestrant *I* 2, 13
Serum phosphate *I* 15
Shabat adaptor unit *I* 68
Sialic acids, influenza virus *I* 39
Sialyltransferase *I* 127
Sickle cell anemia *I* 22
– non-ionic surfactant *I* 52
Signal transduction *II* 179–180
– polymer-coated surfaces affecting
  II 191–192
Silanes
– aminopropyltriethoxysilane *II* 135
– DNA immobilisation *II* 135
Single-force molecular interactions
  II 162
– ligand-receptor interactions
  II 164–166
– mechanical properties of proteins
  II 162–164
SMANCS *II* 111–112
– clinical status *II* 112–113
– quality of life *II* 114–115
Solid tumours, characteristics *II* 106
Starburst dendrimers *I* 142
Starch grains *II* 150–152
Statins *I* 25
Stimuli-triggered drug release *II* 84–85

Stroke *I* 25
Styrene maleic anhydride *II* 7
Styrene-co-maleic
  anhydride-neocarzinostatin *II* 17–18
Super-stealth property *II* 96
Superoxide dismutase *II* 116
Surface-functionalized polymeric micelles
  II 86–87
Synthetic polyelectrolytes *II* 189–191

Taxol *I* 72
Thalassemia *I* 22
Thiol groups, PEGylation *I* 122
Thrombopoiesis *I* 114
Toxins, polymeric sequestrants *I* 32
Transferrin *I* 145
Transglutaminase, PEGylation *I* 125
Triacyl glycerides *I* 48
Tumour cell targeting *II* 9–11

Tumour vasculature, drug targeting
  II 9–11, 42–50
– delivery schedules and vehicles
  II 48–49
– HPMA copolymer-TNP-470 (caplostatin)
  II 46–48
– markers of angiogenesis *II* 43–45
– PEGylated-liposomes *II* 49–50
– peptide motifs *II* 45–46

Vancomycin *I* 32
– resistance *I* 41
Vascular endothelial growth factor *II* 72
Vascular pemeability factor *II* 72
Vaso-occlusion *I* 52
Viral attachment proteins *I* 37
Viral infections *I* 37
Viral proteins *II* 156–157
Viruses, artificial *I* 164

Printing: Krips bv, Meppel
Binding: Stürtz, Würzburg

RETURN TO: **CHEMISTRY LIBRARY**
100 Hildebrand Hall • 510-642-3753

LOAN PERIOD	1	2	3
4		2 HOUR	6

~~ALL BOOKS MAY BE RECALLED AFTER 7 DAYS~~.
~~Renewals may be requested in person~~ using ~~GLADIS~~,
type **inv** ~~followed by your patron ID number.~~

DUE AS STAMPED BELOW.

JUN 1 6 2006

FORM NO. DD 10
3M 5-04

UNIVERSITY OF CALIFORNIA, BERKELEY
Berkeley, California 94720–6000